THE CHEMICAL HISTORY OF A CANDLE

MICHAEL FARADAY

Born at Newington Butts, 22 September, 1791.
Died at Hampton Court, 25 August, 1867.

The Chemical History of a Candle formed the subject of a course of six lectures to children delivered by Faraday at the Royal Institution, where he was Fullerian Professor of Chemistry, during the Christmas holidays of 1860-61. The lectures were printed in the *Chemical News* of the 5th, 12th, 19th, and 26th January, and the 2nd and 9th February, 1861, from shorthand notes taken by Mr. William Crookes, now Sir William Crookes, F.R.S.

The six lectures so reported, with another (on platinum) delivered by Faraday in February 1861, were collected and published in book form in that year by Messrs. Griffin, Bohn & Co., with a preface and notes by Mr. Crookes and with numerous cuts to illustrate the text. Subsequent editions, practically identical, were issued by Messrs. Chatto and Windus.

The present issue, published by arrangement with the latter firm, includes the preface and notes written for the first edition and all the illustrations, but excludes the lecture on platinum, which neither had nor was intended to have any connection with the course of lectures telling *The Chemical History of a Candle.*

B. I.

The Chemical History of a Candle: *a course of lectures delivered by* Michael Faraday

ॐ

With illustrations

ॐ

London ❧ Unit Library, Ltd ❧ 1904

PRINTED AND BOUND BY
HAZELL, WATSON AND VINEY, LD.,
LONDON AND AYLESBURY.

CONTENTS

PAGE

EDITOR'S PREFACE TO THE FIRST EDITION 7

THE LECTURES

I. A CANDLE: THE FLAME — ITS SOURCES — STRUCTURE —
MOBILITY—BRIGHTNESS 9

II. BRIGHTNESS OF THE FLAME—AIR NECESSARY FOR COM-
BUSTION—PRODUCTION OF WATER 27

III. PRODUCTS: WATER FROM THE COMBUSTION—NATURE OF
WATER—A COMPOUND—HYDROGEN 42

IV. HYDROGEN IN THE CANDLE—BURNS INTO WATER—THE
OTHER PART OF WATER—OXYGEN 60

V. OXYGEN PRESENT IN THE AIR—NATURE OF THE ATMO-
SPHERE—ITS PROPERTIES—OTHER PRODUCTS FROM THE
CANDLE—CARBONIC ACID—ITS PROPERTIES . . . 77

VI. CARBON OR CHARCOAL—COAL-GAS—RESPIRATION AND ITS
ANALOGY TO THE BURNING OF A CANDLE—CONCLUSION 96

NOTES 115

EDITOR'S PREFACE

TO THE FIRST EDITION

FROM the primitive pine-torch to the paraffin candle, how wide an interval! between them how vast a contrast! The means adopted by man to illuminate his home at night, stamp at once his position in the scale of civilisation. The fluid bitumen of the far East, blazing in rude vessels of baked earth ; the Etruscan lamp, exquisite in form, yet ill adapted to its office ; the whale, seal, or bear fat, filling the hut of the Esquimaux or Lap with odour rather than light ; the huge wax candle on the glittering altar ; the range of gas lamps in our streets,—all have their stories to tell. All, if they could speak (and, after their own manner, they can), might warm our hearts in telling how they have ministered to man's comfort, love of home, toil, and devotion.

Surely, among the millions of fire-worshippers and fire-users who have passed away in earlier ages, *some* have pondered over the mystery of fire ; perhaps some clear minds have guessed shrewdly near the truth. Think of the time man has lived in hopeless ignorance : think that only during a period which might be spanned by the life of one man, has the truth been known.

Atom by atom, link by link, has the reasoning chain been forged. Some links, too quickly and too slightly made, have given way, and been replaced by better work ; but now the great phenomena are known—the outline is correctly and firmly drawn—cunning artists are filling in the rest, and the child who masters these Lectures knows more of fire than Aristotle did.

The candle itself is now made to light up the dark places of nature ; the blowpipe and the prism are adding to our knowledge of the earth's crust ; but the torch must come first.

Among the readers of this book some few may devote themselves to increasing the stores of knowledge : the Lamp of Science *must* burn. "*Alere flammam.*"

<div align="right">W. CROOKES.</div>

THE CHEMICAL HISTORY

OF A

CANDLE

LECTURE I

A CANDLE : THE FLAME—ITS SOURCES—STRUCTURE—MOBILITY—BRIGHTNESS

I PURPOSE, in return for the honour you do us by coming to see what are our proceedings here, to bring before you, in the course of these lectures, the Chemical History of a Candle. I have taken this subject on a former occasion; and were it left to my own will, I should prefer to repeat it almost every year—so abundant is the interest that attaches itself to the subject, so wonderful are the varieties of outlet which it offers into the various departments of philosophy. There is not a law under which any part of this universe is governed which does not come into play, and is touched upon in these phenomena. There is no better, there is no more open door by which you can enter into the study of natural philosophy, than by considering the physical phenomena of a candle. I trust, therefore, I shall not disappoint you in choosing this for my subject rather than any newer topic, which could not be better, were it even so good.

And before proceeding, let me say this also—that though our subject be so great, and our intention that of treating it honestly, seriously, and philosophically, yet

I mean to pass away from all those who are seniors amongst us. I claim the privilege of speaking to juveniles as a juvenile myself. I have done so on former occasions—and, if you please, I shall do so again. And though I stand here with the knowledge of having the words I utter given to the world, yet that shall not deter me from speaking in the same familiar way to those whom I esteem nearest to me on this occasion.

And now, my boys and girls, I must first tell you of what candles are made. Some are great curiosities. I have here some bits of timber, branches of trees particularly famous for their burning. And here you see a piece of that very curious substance taken out of some of the bogs in Ireland, called *candle-wood*,—a hard, strong, excellent wood, evidently fitted for good work as a resister of force, and yet withal burning so well that where it is found they make splinters of it, and torches, since it burns like a candle, and gives a very good light indeed. And in this wood we have one of the most beautiful illustrations of the general nature of a candle that I can possibly give. The fuel provided, the means of bringing that fuel to the place of chemical action, the regular and gradual supply of air to that place of action—heat and light—all produced by a little piece of wood of this kind, forming, in fact, a natural candle.

But we must speak of candles as they are in commerce. Here are a couple of candles commonly called dips. They are made of lengths of cotton cut off, hung up by a loop, dipped into melted tallow, taken out again and cooled, then re-dipped until there is an accumulation of tallow round the cotton. In order that you may have an idea of the various characters of these candles, you see these which I hold in my hand—they are very small, and very curious. They are, or were, the candles used by the miners in coal mines. In olden times the miner had to find his own candles; and it was supposed that a small candle would not so soon set fire to the firedamp in the coal

10

mines as a large one ; and for that reason, as well as for
economy's sake, he had candles made of this sort—20, 30,
40, or 60 to the pound. They have been replaced since
then by the steel-mill, and then by the Davy-lamp, and
other safety-lamps of various kinds. I have here a candle
that was taken out of the *Royal George*, it is said, by
Colonel Pasley. It has been sunk in the sea for many
years, subject to the action of salt water. It shews you
how well candles may be preserved ; for though it is cracked
about and broken a good deal, yet, when lighted, it goes
on burning regularly, and the tallow resumes its natural
condition as soon as it is fused.

Mr. Field, of Lambeth, has supplied me abundantly
with beautiful illustrations of the candle and its materials.
I shall therefore now refer to them. And, first, there is
the suet—the fat of the ox—Russian tallow, I believe,
employed in the manufacture of these dips, which Gay
Lussac, or some one who entrusted him with his know-
ledge, converted into that beautiful substance, stearin,
which you see lying beside it. A candle, you know, is not
now a greasy thing like an ordinary tallow candle, but
a clean thing, and you may almost scrape off and pulverise
the drops which fall from it without soiling anything.
This is the process he adopted :—The fat or tallow is first
boiled with quicklime, and made into a soap, and then
the soap is decomposed by sulphuric acid, which takes
away the lime, and leaves the fat rearranged as stearic
acid, whilst a quantity of glycerin is produced at the same
time. Glycerin—absolutely a sugar, or a substance similar
to sugar—comes out of the tallow in this chemical change.
The oil is then pressed out of it; and you see here this
series of pressed cakes, shewing how beautifully the im-
purities are carried out by the oily part as the pressure
goes on increasing, and at last you have left that substance
which is melted, and cast into candles as here represented.
The candle I have in my hand is a stearin candle, made
of stearin from tallow in the way I have told you. Then

here is a sperm candle, which comes from the purified oil of the spermaceti whale. Here also are yellow beeswax and refined beeswax, from which candles are made. Here, too, is that curious substance called paraffin, and some paraffin candles made of paraffin obtained from the bogs of Ireland. I have here also a substance brought from Japan, since we have forced an entrance into that out-of-the-way place—a sort of wax which a kind friend has sent me, and which forms a new material for the manufacture of candles.

And how are these candles made? I have told you about dips, and I will show you how moulds are made. Let us imagine any of these candles to be made of materials which can be cast. "Cast!" you say. "Why, a candle is a thing that melts; and surely if you can melt it, you can cast it." Not so. It is wonderful, in the progress of manufacture, and in the consideration of the means best fitted to produce the required result, how things turn up which one would not expect beforehand. Candles cannot always be cast. A wax candle can never be cast. It is made by a particular process, which I can illustrate in a minute or two; but I must not spend much time on it. Wax is a thing which, burning so well, and melting so easily in a candle, cannot be cast. However, let us take a material that can be cast. Here is a frame, with a number of moulds fastened in it. The first thing to be done is to put a wick through them. Here is one—a plaited wick, which does not require snuffing—supported by a little wire. It goes to the bottom, where it is pegged in—the little peg holding the cotton tight, and stopping the aperture, so that nothing fluid shall run out. At the upper part there is a little bar placed across, which stretches the cotton and holds it in the mould. The tallow is then melted, and the moulds are filled. After a certain time, when the moulds are cool, the excess of tallow is poured off at one corner, and then cleaned off altogether, and the ends of the wick cut away. The candles alone then remain

in the mould, and you have only to upset them, as I am doing, when out they tumble, for the candles are made in the form of cones, being narrower at the top than at the bottom ; so that what with their form and their own shrinking, they only need a little shaking, and out they fall. In the same way are made these candles of stearin and of paraffin. It is a curious thing to see how wax candles are made. A lot of cottons are hung upon frames, as you see here, and covered with metal tags at the ends to keep the wax from covering the cotton in those places. These are carried to a heater, where the wax is melted. As you see, the frames can turn round ; and as they turn, a man takes a vessel of wax, and pours it first down one, and then the next, and the next, and so on. When he has gone once round, if it is sufficiently cool, he gives the first a second coat, and so on until they are all of the required thickness. When they have been thus clothed, or fed, or made up to that thickness, they are taken off, and placed elsewhere. I have here, by the kindness of Mr. Field, several specimens of these candles. Here is one only half-finished. They are then taken down, and well rolled upon a fine stone slab, and the conical top is moulded by properly shaped tubes, and the bottoms cut off and trimmed. This is done so beautifully that they can make candles in this way weighing exactly four, or six, to the pound, or any number they please.

We must not, however, take up more time about the mere manufacture, but go a little further into the matter. I have not yet referred you to luxuries in candles (for there is such a thing as luxury in candles). See how beautifully these are coloured : you see here mauve, magenta, and all the chemical colours recently introduced, applied to candles. You observe, also, different forms employed. Here is a fluted pillar most beautifully shaped ; and I have also here some candles sent me by Mr. Pearsall, which are ornamented with designs upon them, so that as they burn you have as it were a glowing sun above, and a bouquet of flowers

beneath. All, however, that is fine and beautiful is not useful. These fluted candles, pretty as they are, are bad candles ; they are bad because of their external shape. Nevertheless, I shew you these specimens sent to me from kind friends on all sides, that you may see what is done, and what may be done in this or that direction ; although, as I have said, when we come to these refinements, we are obliged to sacrifice a little in utility.

Now, as to the light of the candle. We will light one or two, and set them at work in the performance of their proper functions. You observe a candle is a very different thing from a lamp. With a lamp you take a little oil, fill your vessel, put in a little moss or some cotton prepared by artificial means, and then light the top of the wick. When the flame runs down the cotton to the oil, it gets extinguished, but it goes on burning in the part above. Now, I have no doubt you will ask, how is it that the oil, which will not burn of itself, gets up to the top of the cotton, where it will burn? We shall presently examine that ; but there is a much more wonderful thing about the burning of a candle than this. You have here a solid substance with no vessel to contain it ; and how is it that this solid substance can get up to the place where the flame is ? How is it that this solid gets there, it not being a fluid ? or, when it is made a fluid, then how is it that it keeps together ? This is a wonderful thing about a candle.

We have here a good deal of wind, which will help us in some of our illustrations, but tease us in others ; for the sake, therefore, of a little regularity, and to simplify the matter, I shall make a quiet flame—for who can study a subject when there are difficulties in the way not belonging to it ? Here is a clever invention of some costermonger or street stander in the market-place for the shading of their candles on Saturday nights, when they are selling their greens, or potatoes, or fish. I have very often admired it. They put a lamp-glass round the candle, supported on

THE CUP

a kind of gallery, which clasps it, and it can be slipped up and down as required. By the use of this lamp-glass, employed in the same way, you have a steady flame, which you can look at, and carefully examine, as I hope you will do, at home.

You see, then, in the first instance, that a beautiful cup is formed. As the air comes to the candle it moves upwards by the force of the current which the heat of the candle produces, and it so cools all the sides of the wax, tallow, or fuel, as to keep the edge much cooler than the part within; the part within melts by the flame that runs down the wick as far as it can go before it is extinguished, but the part on the outside does not melt. If I made a current in one direction, my cup would be lopsided, and the fluid would consequently run over,—for the same force of gravity which holds worlds together holds this fluid, in a horizontal position, and if the cup be not horizontal, of course the fluid will run away in guttering. You see, therefore, that the cup is formed by this beautifully regular ascending current of air playing upon all sides, which keeps the exterior of the candle cool. No fuel would serve for a candle which has not the property of giving this cup, except such fuel as the Irish bogwood, where the material itself is like a sponge, and holds its own fuel. You see now why you would have had such a bad result if you were to burn these beautiful candles that I have shewn you, which are irregular, intermittent in their shape, and cannot therefore have that nicely formed edge to the cup which is the great beauty in a candle. I hope you will now see that the perfection of a process—that is, its utility—is the better point of beauty about it. It is not the best looking thing, but the best acting thing, which is the most advantageous to us. This good-looking candle is a bad burning one. There will be a guttering round about it because of the irregularity of the stream of air and the badness of the cup which is formed thereby. You may see some pretty examples (and I trust you will notice these instances) of

the action of the ascending current when you have a little gutter run down the side of a candle, making it thicker there than it is elsewhere. As the candle goes on burning, that keeps its place and forms a little pillar sticking up by the side, because, as it rises higher above the rest of the wax or fuel, the air gets better round it, and it is more cooled and better able to resist the action of the heat at a little distance. Now, the greatest mistakes and faults with regard to candles, as in many other things, often bring with them instruction which we should not receive if they had not occurred. We come here to be philosophers; and I hope you will always remember that whenever a result happens, especially if it be new, you should say, "What is the cause? Why does it occur?" and you will in the course of time find out the reason.

Then, there is another point about these candles which will answer a question,—that is, as to the way in which this fluid gets out of the cup, up the wick, and into the place of combustion. You know that the flames on these burning wicks in candles made of beeswax, stearin, or spermaceti, do not run down to the wax or other matter, and melt it all away, but keep to their own right place. They are fenced off from the fluid below, and do not encroach on the cup at the sides. I cannot imagine a more beautiful example than the condition of adjustment under which a candle makes one part subserve to the other to the very end of its action. A combustible thing like that, burning away gradually, never being intruded upon by the flame, is a very beautiful sight; especially when you come to learn what a vigorous thing flame is—what power it has of destroying the wax itself when it gets hold of it, and of disturbing its proper form if it come only too near.

But how does the flame get hold of the fuel? There is a beautiful point about that—*capillary attraction*. "Capillary attraction!" you say,—"the attraction of hairs." Well, never mind the name: it was given in old

16

times, before we had a good understanding of what the real power was. It is by what is called capillary attraction that the fuel is conveyed to the part where combustion goes on, and is deposited there, not in a careless way, but very beautifully in the very midst of the centre of action which takes place around it. Now, I am going to give you one or two instances of capillary attraction. It is that kind of action or attraction which makes two things that do not dissolve in each other still hold together. When you wash your hands, you wet them thoroughly; you take a little soap to make the adhesion better, and

Fig. 1

you find your hand remains wet. This is by that kind of attraction of which I am about to speak. And, what is more, if your hands are not soiled (as they almost always are by the usages of life), if you put your finger into a little warm water, the water will creep a little way up the finger, though you may not stop to examine it. I have here a substance which is rather porous—a column of salt —and I will pour into the plate at the bottom, not water, as it appears, but a saturated solution of salt which cannot absorb more; so that the action which you see will not be due to its dissolving anything. We may consider the plate to be the candle, and the salt the wick, and this solution the melted tallow. (I have coloured the fluid,

that you may see the action better.) You observe that, now I pour in the fluid, it rises and gradually creeps up the salt higher and higher; and provided the column does not tumble over, it will go to the top. If this blue solution were combustible, and we were to place a wick at the top of the salt, it would burn as it entered into the wick. It is a most curious thing to see this kind of action taking place, and to observe how singular some of the circumstances are about it. When you wash your hands, you take a towel to wipe off the water; and it is by that kind of wetting, or that kind of attraction which makes the towel become wet with water, that the wick is made wet with the tallow. I have known some careless boys and girls (indeed, I have known it happen to careful people as well) who, having washed their hands and wiped them with a towel, have thrown the towel over the side of the basin, and before long it has drawn all the water out of the basin and conveyed it to the floor, because it happened to be thrown over the side in such a way as to serve the purpose of a syphon. That you may the better see the way in which the substances act one upon another, I have here a vessel made of wire gauze filled with water, and you may compare it in its action to the cotton in one respect, or to a piece of calico in the other. In fact, wicks are sometimes made of a kind of wire gauze. You will observe that this vessel is a porous thing; for if I pour a little water on to the top, it will run out at the bottom. You would be puzzled for a good while if I asked you what the state of this vessel is, what is inside it, and why it is there? The vessel is full of water, and yet you see the water goes in and runs out as if it were empty. In order to prove this to you, I have only to empty it. The reason is this,—the wire, being once wetted, remains wet; the meshes are so small that the fluid is attracted so strongly from the one side to the other, as to remain in the vessel although it is porous. In like manner the particles of melted tallow ascend the cotton and get to the top; other particles then

follow by their mutual attraction for each other, and as they reach the flame they are gradually burned.

Here is another application of the same principle. You see this bit of cane. I have seen boys about the streets, who are very anxious to appear like men, take a piece of cane, and light it and smoke it, as an imitation of a cigar. They are enabled to do so by the permeability of the cane in one direction, and by its capillarity. If I place this piece of cane on a plate containing some camphine (which is very much like paraffin in its general character), exactly in the same manner as the blue fluid rose through the salt will this fluid rise through the piece of cane. There being no pores at the side, the fluid cannot go in that direction, but must pass through its length. Already the fluid is at the top of the cane : now I can light it and make it serve as a candle. The fluid has risen by the capillary attraction of the piece of cane, just as it does through the cotton in the candle.

Now, the only reason why the candle does not burn all down the side of the wick is, that the melted tallow extinguishes the flame. You know that a candle, if turned upside down, so as to allow the fuel to run upon the wick, will be put out. The reason is, that the flame has not had time to make the fuel hot enough to burn, as it does above, where it is carried in small quantities into the wick, and has all the effect of the heat exercised upon it.

There is another condition which you must learn as regards the candle, without which you would not be able fully to understand the philosophy of it, and that is the vaporous condition of the fuel. In order that you may understand that, let me show you a very pretty, but very common-place experiment. If you blow a candle out cleverly, you will see the vapour rise from it. You have, I know, often smelt the vapour of a blown-out candle— and a very bad smell it is ; but if you blow it out cleverly, you will be able to see pretty well the vapour into which this solid matter is transformed. I will blow out one of

19

these candles in such a way as not to disturb the air around
it, by the continuing action of my breath; and now, if
I hold a lighted taper two or three inches from the wick,
you will observe a train of fire going through the air till
it reaches the candle. I am obliged to be quick and ready,
because, if I allow the vapour time to cool, it becomes
condensed into a liquid or solid, or the stream of com-
bustible matter gets disturbed.

Now, as to the shape or form of the flame. It concerns
us much to know about the condition which the matter
of the candle finally assumes at the top of the wick—where
you have such beauty and brightness as nothing but

Fig. 2

combustion or flame can produce. You have the glittering
beauty of gold and silver, and the still higher lustre of
jewels, like the ruby and diamond; but none of these rival
the brilliancy and beauty of flame. What diamond can
shine like flame? It owes its lustre at night-time to the
very flame shining upon it. The flame shines in darkness,
but the light which the diamond has is as nothing until
the flame shine upon it, when it is brilliant again. The

candle alone shines by itself, and for itself, or for those who have arranged the materials. Now, let us look a little at the form of the flame as you see it under the glass shade. It is steady and equal; and its general form is that which is represented in the diagram, varying with atmospheric disturbances, and also varying according to the size of the candle. It is a bright oblong—brighter at the top than towards the bottom—with the wick in the middle, and besides the wick in the middle, certain darker parts towards the bottom, where the ignition is not so perfect as

Fig. 3

in the part above. I have a drawing here, sketched many years ago by Hooker, when he made his investigations. It is the drawing of the flame of a lamp, but it will apply to the flame of a candle. The cup of the candle is the vessel or lamp, the melted spermaceti is the oil, and the wick is common to both. Upon that he sets this little flame, and then he represents what is true—a certain quantity of matter rising about it which you do not see, and which, if you have not been here before, or are not familiar with the subject, you will not know of. He has here represented the parts of the surrounding atmosphere that are very

essential to the flame, and that are always present with it. There is a current formed, which draws the flame out—for the flame which you see is really drawn out by the current, and drawn upward to a great height—just as Hooker has here shewn you by that prolongation of the current in the diagram. You may see this by taking a lighted candle, and putting it in the sun so as to get its shadow thrown on a piece of paper. How remarkable it is that that thing which is light enough to produce shadows of other objects, can be made to throw its own shadow on a piece of white paper or card, so that you can actually see streaming round the flame something which is not part of the flame, but is ascending and drawing the flame upwards. Now, I am going to imitate the sunlight, by applying the voltaic battery to the electric lamp. You now see our sun, and its great luminosity; and by placing a candle between it and the screen, we get the shadow of the flame. You

Fig. 4

observe the shadow of the candle and of the wick; then there is a darkish part, as represented in the diagram, and then a part which is more distinct. Curiously enough,

however, what we see in the shadow as the darkest part of the flame is, in reality, the brightest part; and here you see streaming upwards the ascending current of hot air, as shewn by Hooker, which draws out the flame, supplies it with air, and cools the sides of the cup of melted fuel.

I can give you here a little further illustration, for the purpose of shewing you how flame goes up or down, according to the current. I have here a flame—it is not a candle flame—but you can, no doubt, by this time, generalise

Fig. 5

enough to be able to compare one thing with another. What I am about to do is to change the ascending current that takes the flame upwards into a descending current. This I can easily do by the little apparatus you see before me. The flame, as I have said, is not a candle flame, but it is produced by alcohol, so that it shall not smoke too much. I will also colour the flame with another substance, so that you may trace its course; for with the spirit alone you could hardly see well enough to have the opportunity of tracing its direction. By lighting this spirit-of-wine, we have then a flame produced; and you observe that

when held in the air, it naturally goes upwards. You understand now easily enough why flames go up under ordinary circumstances—it is because of the draught of air by which the combustion is formed. But now, by blowing the flame down, you see I am enabled to make it go downwards into this little chimney—the direction of the current being changed. Before we have concluded this course of lectures, we shall shew you a lamp in which the flame goes up and the smoke goes down, or the flame goes down and the smoke goes up. You see, then, that we have the power in this way of varying the flame in different directions.

There are now some other points that I must bring before you. Many of the flames you see here vary very much in their shape by the currents of air blowing around them in different directions ; but we can, if we like, make flames so that they will look like fixtures, and we can photograph them—indeed, we have to photograph them—so that they become fixed to us, if we wish to find out everything concerning them. That, however, is not the only thing I wish to mention. If I take a flame sufficiently large, it does not keep that homogeneous, that uniform condition of shape, but it breaks out with a power of life which is quite wonderful. I am about to use another kind of fuel, but one which is truly and fairly a representative of the wax or tallow of a candle. I have here a large ball of cotton, which will serve as a wick. And, now that I have immersed it in spirit and applied a light to it, in what way does it differ from an ordinary candle? Why, it differs very much in one respect, that we have a vivacity and power about it, a beauty and a life entirely different from the light presented by a candle. You see those fine tongues of flame rising up. You have the same general disposition of the mass of the flame from below upwards ; but, in addition to that, you have this remarkable breaking out into tongues which you do not perceive in the case of a candle. Now, why is this? I must explain it to you,

because when you understand that perfectly, you will
be able to follow me better in what I have to say here-
after. I suppose some here will have made for themselves
the experiment I am going to shew you. Am I right in
supposing that anybody here has played at snapdragon?
I do not know a more beautiful illustration of the philo-
sophy of flame, as to a certain part of its history, than the
game of snapdragon. First, here is the dish; and let me
say, that when you play snapdragon properly, you ought
to have the dish well-warmed; you ought also to have
warm plums and warm brandy, which, however, I have not

Fig. 6

got. When you have put the spirit into the dish, you have
the cup and the fuel; and are not the raisins acting like
the wicks? I now throw the plums into the dish, and
light the spirit, and you see those beautiful tongues of
flame that I refer to. You have the air creeping in over
the edge of the dish forming these tongues. Why? Because,
through the force of the current and the irregularity of the
action of the flame, it cannot flow in one uniform stream.
The air flows in so irregularly that you have what would
otherwise be a single image, broken up into a variety of
forms, and each of these little tongues has an independent

existence of its own. Indeed, I might say, you have here a multitude of independent candles. You must not imagine, because you see these tongues all at once, that the flame is of this particular shape. A flame of that shape is never so at any one time. Never is a body of flame, like that which you just saw rising from the ball, of the shape it appears to you. It consists of a multitude of different shapes, succeeding each other so fast that the eye is only able to take cognisance of them all at once. In former times, I purposely analysed a flame of that general character, and the diagram shews you the different parts of which it is composed. They do not occur all at once : it is only because we see these shapes in such rapid succession, that they seem to us to exist all at one time.

It is too bad that we have not got further than my game of snapdragon ; but we must not, under any circumstances, keep you beyond your time. It will be a lesson to me in future to hold you more strictly to the philosophy of the thing, than to take up your time so much with these illustrations.

LECTURE II

WE were occupied the last time we met in considering
the general character and arrangement as regards the fluid
portion of a candle, and the way in which that fluid got
into the place of combustion. You see, when we have a
candle burning fairly in a regular, steady atmosphere, it
will have a shape something like the one shewn in the
diagram, and will look pretty uniform, although very
curious in its character. And now, I have to ask your
attention to the means by which we are enabled to ascertain
what happens in any particular part of the flame—why
it happens, what it does in happening, and where, after
all, the whole candle goes to: because, as you know very
well, a candle being brought before us and burned,
disappears, if burned properly, without the least trace
of dirt in the candlestick—and this is a very curious
circumstance. In order, then, to examine this candle
carefully, I have arranged certain apparatus, the use of
which you will see as I go on. Here is a candle: I am
about to put the end of this glass tube into the middle of
the flame—into that part which old Hooker has represented
in the diagram as being rather dark, and which you can see
at any time, if you will look at a candle carefully, without
blowing it about. We will examine this dark part first.

Now, I take this bent glass tube, and introduce one
end into that part of the flame, and you see at once that
something is coming from the flame, out at the other end

27

of the tube; and if I put a flask there, and leave it for a little while, you will see that something from the middle part of the flame is gradually drawn out, and goes through the tube and into that flask, and there behaves very differently from what it does in the open air. It not only escapes from the end of the tube, but falls down to the bottom of the flask like a heavy substance, as indeed it is. We find that this is the wax of the candle made into a

Fig. 7

vaporous fluid—not a gas. (You must learn the difference between a gas and a vapour: a gas remains permanent, a vapour is something that will condense.) If you blow out a candle, you perceive a very nasty smell, resulting from the condensation of this vapour. That is very different from what you have outside the flame; and, in order to make that more clear to you, I am about to produce and set fire to a larger portion of this vapour—for what we have in the small way in a candle, to under-

stand thoroughly, we must, as philosophers, produce in a larger way, if needful, that we may examine the different parts. And now Mr. Anderson will give me a source of heat, and I am about to shew you what that vapour is. Here is some wax in a glass flask, and I am going to make it hot, as the inside of that candle-flame is hot, and the matter about the wick is hot. [*The Lecturer placed some wax in a glass flask, and heated it over a lamp.*] Now, I dare say that is hot enough for me. You see that the wax I put in it has become fluid, and there is a little smoke coming from it. We shall very soon have

Fig. 8

the vapour rising up. I will make it still hotter, and now we get more of it, so that I can actually pour the vapour out of the flask into that basin, and set it on fire there. This, then, is exactly the same kind of vapour as we have in the middle of the candle; and that you may be sure this is the case, let us try whether we have not got here, in this flask, a real combustible vapour out of the middle of the candle. [*Taking the flask into which the tube from the candle proceeded, and introducing a lighted taper.*] See how it burns. Now, this is the vapour from the middle of the candle, produced by its own heat; and that is one of the first things you have to

consider with respect to the progress of the wax in the course of its combustion, and as regards the changes it undergoes. I will arrange another tube carefully in the flame, and I should not wonder if we were able, by a little care, to get that vapour to pass through the tube to the other extremity, where we will light it, and obtain absolutely the flame of the candle at a place distant from it. Now, look at that. Is not that a very pretty experiment? Talk about laying on gas—why, we can actually lay on a candle! And you see from this that there are clearly two different kinds of action—one the *production* of the vapour, and the other the *combustion* of it—both of which take place in particular parts of the candle.

I shall get no vapour from that part which is already burnt. If I raise the tube to the upper part of the flame, so soon as the vapour has been swept out, what comes away will be no longer combustible: it is already burned. How burned? Why, burned thus :—In the middle of the flame, where the wick is, there is this combustible vapour; on the outside of the flame is the air which we shall find necessary for the burning of the candle; between the two, intense chemical action takes place, whereby the air and the fuel act upon each other, and at the very same time that we obtain light the vapour inside is destroyed. If you examine where the heat of a candle is, you will find it very curiously arranged. Suppose I take this candle, and hold a piece of paper close upon the flame, where is the heat of that flame? Do you not see that it is *not* in the inside? It is in a ring, exactly in the place where I told you the chemical action was; and even in my irregular mode of making the experiment, if there is not too much disturbance, there will always be a ring. This is a good experiment for you to make at home. Take a strip of paper, have the air in the room quiet, and put the piece of paper right across the middle of the flame (I must not talk while I make the experiment), and you will find that it is burnt in two places,

and that it is not burnt, or very little so, in the middle; and when you have tried the experiment once or twice, so as to make it nicely, you will be very interested to see where the heat is, and to find that it is where the air and the fuel come together.

This is most important for us as we proceed with our subject. Air is absolutely necessary for combustion; and, what is more, I must have you understand that *fresh* air is necessary, or else we should be imperfect in our reasoning and our experiments. Here is a jar of air. I place it over a candle, and it burns very nicely in it at first, shewing that what I have said about it is true; but there will soon be a change. See how the flame is drawing upwards, presently fading, and at last going out. And going out, why? Not because it wants air merely, for the jar is as full now as it was before; but it wants pure, fresh air. The jar is full of air, partly changed, partly not changed; but it does not contain sufficient of the fresh air which is necessary for the combustion of a candle. These are all points which we, as young chemists, have to gather up; and if we look a little more closely into this kind of action, we shall find certain steps of reasoning extremely interesting. For instance, here is the oil-lamp I shewed you—an excellent lamp for our experiments—the old Argand lamp. I now make it like a candle [*obstructing the passage of air into the centre of the flame*]; there is the cotton; there is the oil rising up it; and there is the conical flame. It burns poorly, because there is a partial restraint of air. I have allowed no air to get to it, save round the outside of the flame, and it does not burn well. I cannot admit more air from the outside, because the wick is large; but if, as Argand did so cleverly, I open a passage to the middle of the flame, and so let air come in there, you will see how much more beautifully it burns. If I shut the air off, look how it smokes; and why? We have now some very interesting points to study. We have the case of the combustion of a candle; we have the case of a candle being

31

put out by the want of air; and we have now the case of imperfect combustion; and this is to us so interesting, that I want you to understand it as thoroughly as you do the case of a candle burning in its best possible manner. I will now make a great flame, because we need the largest possible illustrations. Here is a larger wick [*burning turpentine on a ball of cotton*]. All these things are the same as candles, after all. If we have larger wicks, we must have a larger supply of air, or we shall have less perfect combustion. Look now at this black substance going up into the atmosphere; there is a regular stream of it. I have provided means to carry off the imperfectly burned part, lest it should annoy you. Look at the soots that fly off from the flame : see what an imperfect combustion it is, because it cannot get enough air. What, then, is happening? Why, certain things which are necessary to the combustion of a candle are absent, and very bad results are accordingly produced ; but we see what happens to a candle when it is burnt in a pure and proper state of air. At the time when I shewed you this charring by the ring of flame on the one side of the paper, I might have also shewn you, by turning to the other side, that the burning of a candle produces the same kind of soot—charcoal or carbon.

But, before I shew that, let me explain to you—as it is quite necessary for our purpose—that, though I take a candle and give you, as the general result, its combustion in the form of a flame, we must see whether combustion is always in this condition, or whether there are other conditions of flame ; and we shall soon discover that there are, and that they are most important to us. I think, perhaps, the best illustration of such a point to us, as juveniles, is to shew the result of strong contrast. Here is a little gunpowder. You know that gunpowder burns with flame—we may fairly call it flame. It contains carbon and other materials, which altogether cause it to burn with a flame. And here is some pulverised iron, or iron-filings.

COMBUSTION WITH AND WITHOUT FLAME

Now, I purpose burning these two things together. I have a little mortar in which I will mix them. (Before I go into these experiments, let me hope that none of you, by trying to repeat them, for fun's sake, will do any harm. These things may all be very properly used if you take care ; but without that, much mischief will be done.) Well, then, here is a little gunpowder, which I put at the bottom of that little wooden vessel, and mix the iron-filings up with it, my object being to make the gunpowder set fire to the filings and burn them in the air, and thereby shew the difference between substances burning with flame and not with flame. Here is the mixture ; and when I set fire to it, you must watch the combustion, and you will see that it is of two kinds. You will see the gunpowder burning with a flame, and the filings thrown up. You will see them burning too, but without the production of flame. They will each burn separately. [*The Lecturer then ignited the mixture.*] There is the gunpowder, which burns with a flame ; and there are the filings—they burn with a different kind of combustion. You see, then, these two great distinctions ; and upon these differences depend all the utility and all the beauty of flame which we use for the purpose of giving out light. When we use oil, or gas, or candle, for the purpose of illumination, their fitness all depends upon these different kinds of combustion.

There are such curious conditions of flame, that it requires some cleverness and nicety of discrimination to distinguish the kinds of combustion one from another. For instance, here is a powder which is very combustible, consisting, as you see, of separate little particles. It is called *lycopodium*, and each of these particles can produce a vapour, and produce its own flame ; but, to see them burning, you would imagine it was all one flame. I will now set fire to a quantity, and you will see the effect. We saw a cloud of flame, apparently in one body ; but that rushing noise [*referring to the sound produced by the burning*] was a proof that the combustion was not a continuous

33 C

or regular one. This is the lightning of the pantomimes, and a very good imitation. [*The experiment was twice repeated by blowing lycopodium from a glass tube through a spirit-flame.*] This is not an example of combustion like that of the filings I have been speaking of, to which we must now return.

Suppose I take a candle, and examine that part of it which appears brightest to our eyes. Why, there I get these black particles, which already you have seen many times evolved from the flame, and which I am now about to evolve in a different way. I will take this candle and clear away the gutterage, which occurs by reason of the currents of air ; and if I now arrange a glass tube so as just to dip into this luminous part, as in our first experiment, only higher, you see the result. In place of having the same white vapour that you had before, you will now have a black vapour. There it goes, as black as ink. It is certainly very different from the white vapour ; and when we put a light to it, we shall find that it does not burn, but that it puts the lights out. Well, these particles, as I said before, are just the smoke of the candle ; and this brings to mind that old employment which Dean Swift recommended to servants for their amusement, namely, writing on the ceiling of a room with a candle. But what is that black substance? Why, it is the same carbon which exists in the candle. How comes it out of the candle? It evidently existed in the candle, or else we should not have had it here. And now I want you to follow me in this explanation. You would hardly think that all those substances which fly about London, in the form of soots and blacks, are the very beauty and life of the flame, and which are burned in it as those iron-filings were burned here. Here is a piece of wire gauze, which will not let the flame go through it ; and I think you will see, almost immediately, that when I bring it low enough to touch that part of the flame which is otherwise so bright, that it quells and quenches it at once, and allows a volume of smoke to rise up.

34

CAUSE OF THE LUMINOSITY OF FLAME

I want you now to follow me in this point,—that whenever a substance burns, as the iron-filings burnt in the flame of gunpowder, without assuming the vaporous state (whether it becomes liquid or remains solid), it becomes exceedingly luminous. I have here taken three or four examples apart from the candle, on purpose to illustrate this point to you ; because what I have to say is applicable to all substances, whether they burn or whether they do not burn,—that they are exceedingly bright if they retain their solid state, and that it is to this presence of solid particles in the candle-flame that it owes its brilliancy.

Here is a platinum-wire, a body which does not change by heat. If I heat it in this flame, see how exceedingly luminous it becomes. I will make the flame dim, for the purpose of giving a little light only, and yet you will see that the heat which it can give to that platinum-wire, though far less than the heat it has itself, is able to raise the platinum-wire to a far higher state of effulgence. This flame has carbon in it ; but I will take one that has no carbon in it. There is a material, a kind of fuel—a vapour, or gas, whichever you like to call it—in that vessel, and it has no solid particles in it ; so I take that because it is an example of flame itself burning without any solid matter whatever ; and if I now put this solid substance in it, you see what an intense heat it has, and how brightly it causes the solid body to glow. This is the pipe through which we convey this particular gas, which we call hydrogen, and which you shall know all about next time we meet. And here is a substance called oxygen, by means of which this hydrogen can burn ; and although we produce, by their mixture, far greater heat than you can obtain from the candle, yet there is very little light. If, however, I take a solid substance, and put that into it, we produce an intense light. If I take a piece of lime, a substance which will not burn, and which will not vaporise by the heat (and because it does not vaporise, remains solid, and

35

remains heated), you will soon observe what happens as to its glowing. I have here a most intense heat, produced by the burning of hydrogen in contact with the oxygen; but there is as yet very little light—not for want of heat, but for want of particles which can retain their solid state; but when I hold this piece of lime in the flame of the hydrogen as it burns in the oxygen, see how it glows! This is the glorious limelight, which rivals the voltaic-light, and which is almost equal to sunlight. I have here a piece of carbon or charcoal, which will burn and give us light exactly in the same manner as if it were burnt as part of a candle. The heat that is in the flame of a candle decomposes the vapour of the wax, and sets free the carbon particles—they rise up heated and glowing as this now glows, and then enter into the air. But the particles when burnt never pass off from a candle in the form of carbon. They go off into the air as a perfectly invisible substance, about which we shall know hereafter.

Is it not beautiful to think that such a process is going on, and that such a dirty thing as charcoal can become so incandescent? You see it comes to this—that all bright flames contain these solid particles; all things that burn and produce solid particles, either during the time they are burning, as in the candle, or immediately after being burnt, as in the case of the gunpowder and iron-filings, —all these things give us this glorious and beautiful light.

I will give you a few illustrations. Here is a piece of phosphorus, which burns with a bright flame. Very well; we may now conclude that phosphorus will produce, either at the moment that it is burning or afterwards, these solid particles. Here is the phosphorus lighted, and I cover it over with this glass for the purpose of keeping in what is produced. What is all that smoke? That smoke consists of those very particles which are produced by the combustion of the phosphorus. Here, again, are two substances. This is chlorate of potassa, and this other sulphuret of

antimony. I shall mix these together a little, and then they may be burnt in many ways. I shall touch them with a drop of sulphuric acid, for the purpose of giving you an illustration of chemical action, and they will instantly burn. [*The Lecturer then ignited the mixture by means of sulphuric acid.*] Now, from the appearance of things, you can judge for yourselves whether they produce solid matter in burning. I have given you the train of reasoning which will enable you to say whether they do or do not;

Fig. 9

for what is this bright flame but the solid particles passing off?

Mr. Anderson has in the furnace a very hot crucible,—I am about to throw into it some zinc-filings, and they will burn with a flame like gunpowder. I make this experiment because you can make it well at home. Now, I want you to see what will be the result of the combustion of this zinc. Here it is burning—burning beautifully like a candle, I may say. But what is all that smoke, and what are those little clouds of wool which will come to you if you cannot come to them, and make themselves sensible to you in the form of the old philosophic wool, as it was called? We shall have left in that crucible, also, a quantity of this woolly matter. But I will take a piece of this same zinc and make an experiment a little more closely at home, as it were. You will have here

the same thing happening. Here is the piece of zinc, there [*pointing to a jet of hydrogen*] is the furnace, and we will set to work and try and burn the metal. It glows, you see : there is the combustion, and there is the white substance into which it burns. And so, if I take that flame of hydrogen as the representative of a candle, and shew you a substance like zinc burning in the flame, you will see that it was merely during the action of combustion that this substance glowed—while it was kept hot ; and if I take a flame of hydrogen, and put this white substance from the zinc into it, look how beautifully it glows, and just because it is a solid substance.

I will now take such a flame as I had a moment since, and set free from it the particles of carbon. Here is some camphine, which will burn with a smoke ; but if I send these particles of smoke through this pipe into the hydrogen flame, you will see they will burn and become luminous, because we heat them a second time. There they are. Those are the particles of carbon re-ignited a second time. They are those particles which you can easily see by holding a piece of paper behind them, and which, whilst they are in the flame, are ignited by the heat produced, and, when so ignited, produce this brightness. When the particles are not separated, you get no brightness. The flame of coal-gas owes its brightness to the separation, during combustion, of these particles of carbon, which are equally in that as in a candle. I can very quickly alter that arrangement. Here, for instance, is a bright flame of gas. Supposing I add so much air to the flame as to cause it all to burn before those particles are set free, I shall not have this brightness ; and I can do that in in this way :—If I place over the jet this wire-gauze cap, as you see, and then light the gas over it, it burns with a non-luminous flame, owing to its having plenty of air mixed with it before it burns ; and if I raise the gauze, you see it does not burn below. There is plenty of carbon in the gas ; but, because the atmosphere can get to it, and

mix with it before it burns, you see how pale and blue the flame is. And if I blow upon a bright gas-flame, so as to consume all this carbon before it gets heated to the glowing point, it will also burn blue. [*The Lecturer illustrated his remarks by blowing on the gaslight.*] The only reason why I have not the same bright light when I thus blow upon the flame is, that the carbon meets with sufficient air to burn it before it gets separated in the flame in a free state. The difference is solely due to the solid particles not being separated before the gas is burnt.

You observe that there are certain products as the result

Fig. 10

of the combustion of a candle, and that of these products one portion may be considered as charcoal, or soot; that charcoal, when afterwards burnt, produces some other

product ; and it concerns us very much now to ascertain what that other product is. We shewed that something was going away ; and I want you now to understand how much is going up into the air ; and for that purpose we will have combustion on a little larger scale. From that candle ascends heated air, and two or three experiments will shew you the ascending current ; but, in order to give you a notion of the quantity of matter which ascends in this way, I will make an experiment by which I shall try to imprison some of the products of this combustion. For this purpose I have here what boys call a fire-balloon. I use this fire-balloon merely as a sort of measure of the result of the combustion we are considering ; and I am about to make a flame in such an easy and simple manner as shall best serve my present purpose. This plate shall be the "cup," we will so say, of the candle ; this spirit shall be our fuel ; and I am about to place this chimney over it, because it is better for me to do so than to let things proceed at random. Mr. Anderson will now light the fuel, and here at the top we shall get the results of the combustion. What we get at the top of that tube is exactly the same, generally speaking, as you get from the combustion of a candle ; but we do not get a luminous flame here, because we use a substance which is feeble in carbon. I am about to put this balloon—not into action, because that is not my object—but to shew you the effect which results from the action of those products which arise from the candle, as they arise here from the furnace. [*The balloon was held over the chimney, when it immediately commenced to fill.*] You see how it is disposed to ascend ; but we must not let it up, because it might come in contact with those upper gaslights, and that would be very inconvenient. [*The upper gaslights were turned out, at the request of the Lecturer, and the balloon was allowed to ascend.*] Does not that shew you what a large bulk of matter is being evolved ? Now, there is going through this tube [*placing*

40

a large glass tube over a candle] all the products of that candle, and you will presently see that the tube will become quite opaque. Suppose I take another candle, and place it under a jar, and then put a light on the other side, just to shew you what is going on. You see that the sides of the jar become cloudy, and the light begins to burn feebly. It is the products, you see, which make the light so dim, and this is the same thing which makes the sides of the jar so opaque. If you go home and take a spoon that has been in the cold air, and hold it over a candle— not so as to soot it—you will find that it becomes dim, just as that jar is dim. If you can get a silver dish, or something of that kind, you will make the experiment still better. And now, just to carry your thoughts forward to the time we shall next meet, let me tell you that it is *water* which causes the dimness; and when we next meet, I will show you that we can make it, without difficulty, assume the form of a liquid.

LECTURE III

PRODUCTS : WATER FROM THE COMBUSTION—NATURE OF WATER—A COMPOUND—HYDROGEN

I DARE say you will remember that when we parted we had just mentioned the word " products " from the candle. For when a candle burns we found we were able, by nice adjustment, to get various products from it. There was one substance which was not obtained when the candle was burning properly, which was charcoal or smoke ; and there was some other substance that went upwards from the flame which did not appear as smoke, but took some other form, and made part of that general current which, ascending from the candle upwards, becomes invisible, and escapes. There were also other products to mention. You remember that in that rising current having its origin at the candle, we found that one part was condensable against a cold spoon, or against a clean plate or any other cold thing, and another part was incondensable.

We will first take the condensable part, and examine it ; and, strange to say, we find that that part of the product is just water—nothing but water. On the last occasion I spoke of it incidentally, merely saying that water was produced among the condensable products of the candle ; but to-day I wish to draw your attention to water, that we may examine it carefully, especially in relation to this subject, and also with respect to its general existence on the surface of the globe.

Now, having previously arranged an experiment for the purpose of condensing water from the products of the

candle, my next point will be to shew you this water; and perhaps one of the best means that I can adopt for shewing its presence to so many at once, is to exhibit a very visible action of water, and then to apply that test to what is collected as a drop at the bottom of that vessel. I have here a chemical substance, discovered by Sir Humphrey Davy, which has a very energetic action upon water, which I shall use as a test of the presence of water. If I take a little piece of it—it is called potassium, as coming from potash,—if I take a little piece of it, and throw it into that basin, you see how it shews the presence of water by lighting up and floating about, burning with a violent flame. I am now going to take away the candle which has been burning beneath the vessel containing ice and salt, and you see a drop of water—a condensed product

Fig. 11

of the candle—hanging from under the surface of the dish. I will shew you that potassium has the same action upon it as upon the water in that basin in the experiment we have just tried. See, it takes fire, and burns in just the

same manner. I will take another drop upon this glass slab, and when I put the potassium on to it, you see at once, from its taking fire, that there is water present. Now, that water was produced by the candle. In the same manner, if I put this spirit-lamp under that jar, you will soon see the latter become damp, from the dew which is deposited upon it—that dew being the result of combustion; and I have no doubt you will shortly see by the drops of water which fall upon the paper below, that there is a good deal of water produced from the combustion of the lamp. I will let it remain, and you can afterwards see how much water has been collected. So, if I take a gas-lamp, and put any cooling arrangement over it, I shall get water—water being likewise produced from the combustion of gas. Here, in this bottle, is a quantity of water—perfectly pure, distilled water, produced from the combustion of a gas-lamp—in no point different from the water that you distil from the river, or ocean, or spring, but exactly the same thing. Water is one individual thing—it never changes. We can add to it by careful adjustment, for a little while, or we can take it apart, and get other things from it; but water, as water, remains always the same, either in a solid, liquid, or fluid state. Here, again [*holding another bottle*], is some water produced by the combustion of an oil-lamp. A pint of oil, when burnt fairly and properly, produces rather more than a pint of water. Here, again, is some water, produced by a rather long experiment from a wax candle. And so we can go on with almost all combustible substances, and find that if they burn with a flame, as a candle, they produce water. You may make these experiments yourselves. The head of a poker is a very good thing to try with, and if it remains cold long enough over the candle, you may get water condensed in drops on it; or a spoon or ladle, or anything else may be used, provided it be clean, and can carry off the heat and so condense the water.

And now—to go into the history of this wonderful pro-

duction of water from combustibles, and by combustion—I must first of all tell you that this water may exist in different conditions; and although you may now be acquainted with all its forms, they still require us to give a little attention to them for the present, so that we may perceive how the water, whilst it goes through its Protean changes, is entirely and absolutely the same thing, whether it is produced from a candle, by combustion, or from the rivers or ocean.

First of all, water, when at the coldest, is ice. Now, we philosophers—I hope that I may class you and myself together in this case—speak of water as water, whether it be in its solid, or liquid, or gaseous state,—we speak of it chemically as water. Water is a thing compounded of two substances, one of which we have derived from the candle, and the other we shall find elsewhere. Water may occur as ice; and you have had most excellent opportunities lately of seeing this. Ice changes back into water—for we had on our last Sabbath a strong instance of this change, by the sad catastrophe which occurred in our own house, as well as in the houses of many of our friends,—ice changes back into water when the temperature is raised: water also changes into steam when it is warmed enough. The water which we have here before us is in its densest state, and although it changes in weight, in condition, in form, and in many other qualities, it still is water; and whether we alter it into ice by cooling, or whether we change it into steam by heat, it increases in volume,—in the one case very strangely and powerfully, and in the other case very largely and wonderfully. For instance, I will now take this tin cylinder, and pour a little water into it; and seeing how much water I pour in, you may easily estimate for yourselves how high it will rise in the vessel: it will cover the bottom about two inches. I am now about to convert the water into steam, for the purpose of shewing to you the different volumes which water occupies in its different states of water and steam.

Let us now take the case of water changing into ice : we can effect that by cooling it in a mixture of salt and pounded ice ; and I shall do so to shew you the expansion of water into a thing of larger bulk when it is so changed. These bottles [*holding one*] are made of strong cast iron, very strong and very thick—I suppose they are the third of an inch in thickness ; they are very carefully filled with water, so as to exclude all air, and then they are screwed down tight. We shall see that when we freeze the water in these iron vessels, they will not be able to hold the ice, and the expansion within them will break them in pieces as these [*pointing to some fragments*] are broken, which have been bottles of exactly the same kind. I am about to put these two bottles into that mixture of ice and salt, for the purpose of shewing that when water becomes ice, it changes in volume in this extraordinary way.

In the meantime look at the change which has taken place in the water to which we have applied heat—it is losing its fluid state. You may tell this by two or three circumstances. I have covered the mouth of this glass flask, in which water is boiling, with a watch-glass. Do you see what happens? It rattles away like a valve chattering, because the steam rising from the boiling water sends the valve up and down, and forces itself out, and so makes it clatter. You can very easily perceive that the flask is quite full of steam, or else it would not force its way out. You see, also, that the flask contains a substance very much larger than the water, for it fills the whole of the flask over and over again, and there it is blowing away into the air ; and yet you cannot observe any great diminution in the bulk of the water, which shews you that its change of bulk is very great when it becomes steam.

I have put our iron bottles containing water into this freezing mixture, that you may see what happens. No communication will take place, you observe, between the water in the bottles and the ice in the outer vessel. But

there will be a conveyance of heat from the one to the other; and if we are successful—we are making our experiment in very great haste—I expect you will by-and-by, so soon as the cold has taken possession of the bottles and their contents, hear a pop on the occasion of the bursting of the one bottle or the other; and, when we come to examine the bottles, we shall find their contents masses of ice, partly enclosed by the covering of iron which is too small for them, because the ice is larger in bulk than the water. You know very well that ice floats upon water: if a boy falls through a hole into the water, he tries to get on the ice again to float him up. Why does the ice float?—think of that, and philosophise. Because the ice is larger than the quantity of water which can produce it; and therefore the ice weighs the lighter, and the water is the heavier.

To return now to the action of heat on water. See what a stream of vapour is issuing from this tin vessel! You observe, we must have made it quite full of steam to have it sent out in that great quantity. And now, as we can convert the water into steam by heat, we convert it back into liquid water by the application of cold. And if we take a glass, or any other cold thing, and hold it over this steam, see how soon it gets damp with water; it will condense it until the glass is warm—it condenses the water which is now running down the sides of it. I have here another experiment to shew the condensation of water from a vaporous state back into a liquid state, in the same way as the vapour, one of the products of the candle, was condensed against the bottom of the dish, and obtained in the form of water; and to shew you how truly and thoroughly these changes take place, I will take this tin flask, which is now full of steam, and close the top. We shall see what takes place when we cause this water or steam to return back to the fluid state by pouring some cold water on the outside. [*The Lecturer poured the cold water over the vessel, when it immediately collapsed.*] You

see what has happened. If I had closed the stopper, and still kept the heat applied to it, it would have burst the vessel; yet, when the steam returns to the state of water, the vessel collapses, there being a vacuum produced inside by the condensation of the steam. I shew you these experi-

Fig. 12

ments for the purpose of pointing out that in all these occurrences there is nothing that changes the water into any other thing—it still remains water; and so the vessel is obliged to give way, and is crushed inwards, as in the other case, by the further application of heat, it would have been blown outwards.

And what do you think the bulk of that water is when it assumes the vaporous condition? You see that cube [*pointing to a cubic foot.*] There, by its side, is a cubic inch, exactly the same shape as the cubic foot, and that bulk of water

[*the cubic inch*] is sufficient to expand into that bulk [*the cubic foot*] of steam ; and, on the contrary, the application of cold will contract that large quantity of steam into this small quantity of water. [*One of the iron bottles burst at that moment.*] Ah ! There is one of our bottles burst, and here you see is a crack down one side an eighth of an inch in width. [*The other now exploded, sending the freezing mixture in all directions.*] This other bottle is also broken ; although the iron was nearly half-an-inch thick, the ice has burst it asunder. These changes always take place in water : they do not require to be always produced by artificial means,—we only use them here because we want to produce a small winter round that little bottle, instead of

Fig. 13

a long and severe one. But if you go to Canada, or to the North, you will find the temperature there out of doors will do the same thing as has been done here by the freezing mixture.

To return to our quiet philosophy. We shall not in future be deceived, therefore, by any changes that are produced in water. Water is the same everywhere, whether produced from the ocean or from the flame of the candle. Where, then, is this water which we get from a candle ? I must anticipate a little, and tell you. It evidently comes, as to part of it, from the candle ; but is it within the candle beforehand ? No. It is not in the candle ; and it is not in the air round about the candle which is necessary for its combustion. It is neither in one nor the other, but it

comes from their conjoint action, a part from the candle, a part from the air; and this we have now to trace, so that we may understand thoroughly what is the chemical history of a candle when we have it burning on our table. How shall we get at this? I myself know plenty of ways, but I want *you* to get at it from the association in your own minds of what I have already told you.

I think you can see a little in this way. We had just now the case of a substance which acted upon the water in the way that Sir Humphrey Davy shewed us, and which I am now going to recall to your minds by making again an experiment upon that dish. It is a thing which we have to handle very carefully, for you see, if I allow a little splash of water to come upon this mass, it sets fire to part of it; and if there were free access of air, it would quickly set fire to the whole. Now, this is a metal—a beautiful and bright metal—which rapidly changes in the air, and, as you know, rapidly changes in water. I will put a piece on the water, and you see it burns beautifully, making a floating lamp, using the water in the place of air. Again, if we take a few iron-filings or turnings, and put them in water, we find that they likewise undergo an alteration. They do not change so much as this potassium does, but they change somewhat in the same way; they become rusty, and shew an action upon the water, though in a different degree of intensity to what this beautiful metal does: but they act upon the water in the same manner generally as this potassium. I want you to put these different facts together in your minds. I have another metal here [*zinc*], and when we examined it with regard to the solid substance produced by its combustion, we had an opportunity of seeing that it burned; and I suppose, if I take a little strip of this zinc and put it over the candle, you will see something half-way, as it were, between the combustion of potassium on the water and the action of iron,—you see there is a sort of combustion. It has burned, leaving a white ash or residuum, and here also

we find that the metal has a certain amount of action upon water.

By degrees we have learned how to modify the action of these different substances, and to make them tell us what we want to know. And now, first of all, I take iron. It is a common thing in all chemical reactions, where we get any result of this kind, to find that it is increased by the action of heat; and if we want to examine minutely and carefully the action of bodies one upon another, we often have to refer to the action of heat. You are aware, I believe, that iron-filings burn beautifully in the air; but I am about to show you an experiment of this kind, because it will impress upon you what I am going to say about iron in its action on water. If I take a flame and make it hollow;—you know why, because I want to get air to it and into it, and therefore I make it hollow—and then take a few iron-filings and drop them into the flame, you see how well they burn. That combustion results from the chemical action which is going on when we ignite those particles. And so we proceed to consider these different effects, and ascertain what iron will do when it meets with water. It will tell us the story so beautifully, so gradually and regularly, that I think it will please you very much.

I have here a furnace with a pipe going through it like an iron gun-barrel, and I have stuffed that barrel full of bright iron-turnings, and placed it across the fire, to be made red-hot. We can either send air through the barrel to come in contact with the iron, or we can send steam from this little boiler at the end of the barrel. Here is a stopcock which shuts off the steam from the barrel until we wish to admit it. There is some water in these glass jars, which I have coloured blue, so that you may see what happens. Now, you know very well that any steam I might send through that barrel, if it went through into the water, would be condensed; for you have seen that steam cannot retain its gaseous form if it be

cooled down. You saw it here [*pointing to the tin flask*] crushing itself into a small bulk, and causing the flask holding it to collapse; so that if I were to send steam through that barrel, it would be condensed—supposing the

Fig. 14

barrel were cold : it is, therefore, heated to perform the experiment I am now about to shew you. I am going to send the steam through the barrel in small quantities ; and you shall judge for yourselves, when you see it issue from the other end, whether it still remains steam. Steam is condensable into water, and when you lower the temperature of steam, you convert it back into fluid water ; but I have lowered the temperature of the gas which I have collected in this jar, by passing it through water after it has traversed the iron barrel, and still it does not change back into water. I will take another test and apply to this gas. (I hold the jar in an inverted position, or my substance would escape.) If I now apply a light to the mouth of the jar, it ignites with a slight noise. That tells you that it is not steam. Steam puts out a fire— it does not burn ; but you saw that what I had in that jar burnt. We may obtain this substance equally from

water produced from the candle-flame as from any other source. When it is obtained by the action of the iron upon the aqueous vapour, it leaves the iron in a state very similar to that in which these filings were after they were burnt. It makes the iron heavier than it was before. So long as the iron remains in the tube and is heated, and is cooled again without the access of air or water, it does not change in its weight; but after having had this current of steam passed over it, it then comes out heavier than it was before, having taken something out of the steam, and having allowed something else to pass forth, which we see here. And now, as we have another jar full, I will shew you something most interesting. It is a combustible gas; and I might at once take this jar and set fire to the contents, and shew you that it is

Fig. 15

combustible; but I intend to shew you more if I can. It is also a very light substance. Steam will condense: this body will rise in the air, and not condense. Suppose I take another glass jar, empty of all but air: if I examine it with a taper, I shall find that it contains nothing but air. I will now take this jar full of the gas that I am speaking of, and deal with it as though it were a light body. I will hold both upside-down, and turn the one up under the other; and that which did contain the gas procured from the steam, what does it contain now? You will find it now only contains air. But look! Here is

the combustible substance [*taking the other jar*] which I have poured out of the one jar into the other. It still preserves its quality, and condition, and independence, and therefore is the more worthy of our consideration, as belonging to the products of a candle.

Now, this substance which we have just prepared by the action of iron on the steam or water, we can also get by means of those other things which you have already seen act so well upon the water. If I take a piece of potassium, and make the necessary arrangements, it will produce this gas ; and if, instead, a piece of zinc, I find, when I come to examine it very carefully, that the main reason why this zinc cannot act upon the water continuously as the other metal does, is because the result of the action of the water envelops the zinc in a kind of protecting coat. We have learned in consequence, that if we put into our vessel only the zinc and water, they by themselves do not give rise to much action, and we get no result. But suppose I proceed to dissolve off this varnish—this encumbering substance—which I can do by a little acid ; the moment I do this, I find the zinc acting upon the water exactly as the iron did, but at the common temperature. The acid in no way is altered, except in its combination with the oxide of zinc, which is produced. I have now poured the acid into the glass, and the effect is as though I were applying heat to cause this boiling up. There is something coming off from the zinc very abundantly, which is not steam. There is a jar full of it ; and you will find that I have exactly the same combustible substance remaining in the vessel, when I hold it upside-down, that I produced during the experiment with the iron barrel. This is what we get from water—the same substance which is contained in the candle.

Let us now trace distinctly the connection between these two points. This is hydrogen—a body classed among those things which in Chemistry we call elements, because we can get nothing else out of them. A candle is not an elementary

body, because we can get carbon out of it ; we can get this hydrogen out of it, or at least out of the water which it supplies. And this gas has been so named hydrogen,

Fig. 16

because it is that element which, in association with another, generates water.[1] Mr. Anderson having now been able to get two or three jars of gas, we shall have a few experiments to make, and I want to shew you the best way of making these experiments. I am not afraid to shew you, for I wish you to make experiments, if you will only make them with care and attention, and the assent of those around you. As we advance in Chemistry, we are obliged to deal with substances which are rather injurious, if in their wrong places—the acids, and heat, and combustible things we use, might do harm if carelessly employed. If you want to make hydrogen, you can make it easily from bits of zinc, and sulphuric or muriatic acid. Here is what in former times was called the "philosopher's candle." It is a little phial with a cork, and a tube or pipe passing through it. And I am now putting a few little pieces of zinc into it. This little instrument I am going to apply to a useful purpose in our demonstrations— for I want to shew you that you can prepare hydrogen, and make some experiments with it as you please at your own homes. Let me here tell you why I am so careful

[1] Ύδωρ, "water," and γεννάω, "I generate."

55

to fill this phial nearly, and yet not quite full. I do it because the evolved gas, which, as you have seen, is very combustible, is explosive to a considerable extent when mixed with air, and might lead to harm, if you were to

Fig. 17

apply a light to the end of that pipe before all the air had been swept out of the space above the water. I am now about to pour in the sulphuric acid. I have used very little zinc, and more sulphuric acid and water, because I want to keep it at work for some time. I therefore take care in this way to modify the proportions of the ingredients, so that I may have a regular supply—not too quick, and not too slow. Supposing I now take a glass and put it upside-down over the end of the tube, because the hydrogen is light I expect that it will remain in that vessel a little while. We will now test the contents of our glass to see if there be hydrogen in it. I think I am safe in saying we have caught some [*applying a light*]. There it is you see. I will now apply a light to the top of the tube. There is the hydrogen burning. There is our philosophical candle. It is a foolish feeble sort of a flame, you may say ; but it is so hot that scarcely any common flame gives out so much heat. It goes on burning regularly, and I am

now about to put that flame to burn under a certain arrangement, in order that we may examine its results and make use of the information which we may thereby acquire. Inasmuch as the candle produces water, and this gas comes out of the water, let us see what this gives us by the same process of combustion that the candle went through when it burnt in the atmosphere; and for that purpose I am going to put the lamp under this apparatus, in order to condense whatever may arise from the combustion within it. In the course of a short time you will see moisture appearing in the cylinder, and you will get the water running down the side; and the water from this hydrogen flame will have absolutely the same effect upon all our tests, being obtained by the same general process as in the former case. This hydrogen is a very beautiful substance. It is so light that it carries things up: it is far lighter than the atmosphere; and I dare say I can shew you this by an experiment which, if you are very clever, some of you may even have skill enough to repeat. Here is our generator of hydrogen, and here are some soapsuds. I have an india-rubber tube connected with the hydrogen generator, and at the end

Fig. 18

of the tube is a tobacco-pipe. I can thus put the pipe into the suds, and blow bubbles by means of the hydrogen.

You observe how the bubbles fall downwards when I blow them with my warm breath; but notice the difference when I blow them with hydrogen. [*The Lecturer here blew bubbles with hydrogen, which rose to the roof of the theatre.*] It shews you how light this gas must be in order to carry with it not merely the ordinary soap-bubble, but the larger portion of a drop hanging to the bottom of it. I can shew its lightness in a better way than this; larger bubbles than these may be so lifted up; indeed, in former times balloons used to be filled with this gas. Mr. Anderson will fasten this tube on to our generator, and we shall have a stream of hydrogen here with which we can charge this balloon made of collodion. I need not even be very careful to get all the air out, for I know the power of this gas to carry it up. [*Two collodion balloons were inflated, and sent up, one being held by a string.*] Here is another larger one made of thin membrane, which we will fill and allow to ascend. You will see they will all remain floating about until the gas escapes.

What, then, are the comparative weights of these substances? I have a table here which will shew you the proportion which their weights bear to each other. I have taken a pint and a cubic foot as the measures, and have placed opposite to them the respective figures. A pint measure of this hydrogen weighs three-quarters of our smallest weight (a grain), and a cubic foot weighs one-twelfth of an ounce; whereas a pint of water weighs 8,750 grains, and a cubic foot of water weighs almost 1,000 ounces. You see, therefore, what a vast difference there is between the weight of a cubic foot of water and a cubic foot of hydrogen.

Hydrogen gives rise to no substance that can become solid, either during combustion or afterwards as a product of its combustion. But when it burns, it produces water only; and if we take a cold glass and put it over the flame, it becomes damp, and you have water produced immediately in appreciable quantity; and nothing is pro-

duced by its combustion but the same water which you have seen the flame of the candle produce. It is important to remember that this hydrogen is the only thing in nature which furnishes water as the sole product of combustion.

And now we must endeavour to find some additional proof of the general character and composition of water; and for this purpose I will keep you a little longer, so that at our next meeting we may be better prepared for the subject. We have the power of arranging the zinc which you have seen acting upon the water by the assistance of an acid, in such a manner as to cause all the power to be evolved in the place where we require it. I have behind me a voltaic pile, and I am just about to shew you, at the end of this lecture, its character and power, that you may see what we shall have to deal with when next we meet. I hold here the extremities of the wires which transport the power from behind me, and which I shall cause to act on the water.

We have previously seen what a power of combustion is possessed by the potassium, or the zinc, or the iron-filings; but none of them shew such energy as this. [*The Lecturer here made contact between the two terminal wires of the battery, when a brilliant flash of light was produced.*] This light is, in fact, produced by a forty-zinc power of burning: it is a power that I can carry about in my hands, through these wires, at pleasure—although, if I applied it wrongly to myself, it would destroy me in an instant, for it is a most intense thing, and the power you see here put forth while you count five [*bringing the poles in contact, and exhibiting the electric light*] is equivalent to the power of several thunderstorms, so great is its force. And that you may see what intense energy it has, I will take the ends of the wires which convey the power from the battery, and with it I dare say I can burn this iron file. Now, this is a chemical power, and one which, when we next meet, I shall apply to water, and shew you what results we are able to produce.

LECTURE IV

HYDROGEN IN THE CANDLE—BURNS INTO WATER—THE
OTHER PART OF WATER—OXYGEN

I SEE you are not tired of the candle yet, or I am sure
you would not be interested in the subject in the way you
are. When our candle was burning, we found it produced
water exactly like the water we have around us; and by
further examination of this water we found in it that
curious body, hydrogen—that light substance of which there
is some in this jar. We afterwards saw the burning powers
of that hydrogen, and that it produced water. And I think
I introduced to your notice an apparatus which I very
briefly said was an arrangement of chemical force, or power,
or energy, so adjusted as to convey its power to us in these
wires; and I said I should use that force to pull the water
to pieces, to see what else there was in the water besides
hydrogen; because, you remember, when we passed the
water through the iron tube, we by no means got the
weight of water back which we put in, in the form of steam,
though we had a very large quantity of gas evolved. We
have now to see what is the other substance present. That
you may understand the character and use of this instru-
ment, let us make an experiment or two. Let us put
together, first of all, some substances, knowing what they
are, and then see what that instrument does to them.
There is some copper (observe the various changes which
it can undergo), and here is some nitric acid, and you
will find that this, being a strong chemical agent, will act
very powerfully when I add it to the copper. It is now

60

sending forth a beautiful red vapour; but as we do not want that vapour, Mr. Anderson will hold it near the chimney for a short time, that we may have the use and beauty of the experiment without the annoyance. The copper which I have put into the flask will dissolve: it will change the acid and the water into a blue fluid, containing copper and other things; and I propose then shewing you how this voltaic battery deals with it; and in the meantime we will arrange another kind of experiment for you to see what power it has. This is a substance which is to us like water—that is to say, it contains bodies which we do not know of as yet, as water contains a body which we do not know as yet. Now, this solution of a salt I will put upon paper, and spread about, and apply the power of the battery to it, and observe what will happen. Three or four important things will happen which we shall take advantage of. I place this wetted paper upon a sheet of tinfoil, which is convenient for keeping all clean, and also for the advantageous application of the power; and this solution, you see, is not at all affected by being put upon the paper or tinfoil, nor by anything else I have brought in contact with it yet, and which, therefore, is free to us to use as regards that instrument. But first let us see that our instrument is in order. Here are our wires. Let us see whether it is in the state in which it was last time. We can soon tell. As yet, when I bring them together, we have no power, because the conveyers—what we call the electrodes—the passages or ways for the electricity—are stopped; but now Mr. Anderson by that [*referring to a sudden flash at the ends of the wires*] has given me a telegram to say that it is ready. Before I begin our experiment I will get Mr. Anderson to break contact again at the battery behind me, and we will put a platinum-wire across to connect the poles, and then if I find I can ignite a pretty good length of this wire, we shall be safe in our experiment. Now you will see the power. [*The connection was established, and the*

intermediate wire became red-hot.] There is the power running beautifully through the wire, which I have made thin on purpose to shew you that we have those powerful forces; and now, having that power, we will proceed with it to the examination of water.

I have here two pieces of platinum, and if I lay them down upon this piece of paper [*the moistened paper on the tinfoil*], you will see no action; and if I take them up, there is no change that you can see, but the arrangement remains just as it was before. But, now, see what happens: if I take these two poles and put either one or the other of them down separately on the platinum-plates, they do nothing for me, both are perfectly without action; but if I let them both be in contact at the same moment, see what happens [*a brown spot appeared under each pole of the battery.*] Look here at the effect that takes place, and see how I have pulled something apart from the white—something brown; and I have no doubt, if I were to arrange it thus, and were to put one of the poles to the tinfoil on the other side of the paper—why, I get such a beautiful action upon the paper, that I am going to see whether I cannot write with it—a telegram, if you please. [*The Lecturer here traced the word "juvenile" on the paper with one of the terminal wires.*] See how beautifully we can get our results!

You see we have here drawn something, which we have not known about before, out of this solution. Let us now take that flask from Mr. Anderson's hands, and see what we can draw out of that. This, you know, is a liquid which we have just made up from copper and nitric acid, whilst our other experiments were in hand; and though I am making this experiment very hastily, and may bungle a little, yet I prefer to let you see what I do, rather than prepare it beforehand.

Now, see what happens. These two platinum-plates are the two ends (or I will make them so immediately) of this apparatus; and I am about to put them in contact with

that solution just as we did a moment ago on the paper. It does not matter to us whether the solution be on the paper or whether it be in the jar, so long as we bring the ends of the apparatus to it. If I put the two platinums in by themselves, they come out as clean and as white as they go in [*inserting them into the fluid without connecting them with the battery*]; but when we take the power and lay that on [*the platinums were connected with the battery and again dipped into the solution*], this, you see [*exhibiting one of the platinums*], is at once turned into copper, as it were: it has become like a plate of copper; and that [*exhibiting the other piece of platinum*] has come out quite clean. If I take this coppered piece and change sides, the copper will leave the right-hand side and come over to the left side; what was before the coppered plate comes out clean, and the plate which was clean comes out coated with copper; and thus you see that the same copper we put into this solution we can also take out of it by means of this instrument.

Putting that solution aside, let us now see what effect this instrument will have upon water. Here are two little platinum-plates which I intend to make the ends of the battery, and this (c) is a little vessel so shaped as to enable me to take it to pieces, and shew you its construction. In those two cups (A and B) I pour mercury, which touches the ends of the wires connected with the platinum-plates. In the vessel (c) I pour some water containing a little acid (but which is put only for the purpose of facilitating the action; it undergoes no change in the process), and connected with the top of the vessel is a bent glass tube (D), which may remind you of the pipe which was connected with the gun barrel in our furnace experiment, and which now passes under the jar (F). I have now adjusted this apparatus, and we will proceed to affect the water in some way or other. In the other case, I sent the water through a tube which was made red-hot; I am now going to pass the electricity through the contents of this vessel. Perhaps

I may boil the water; if I do boil the water, I shall get steam; and you know that steam condenses when it gets cold, and you will therefore see by that whether I do boil the water or not. Perhaps, however, I shall not boil the water, but produce some other effect. You shall have the experiment and see. There is one wire which I will put to this side (A), and here is the other wire which I will put to the other side (B), and you will soon see whether any disturbance takes place. Here it is seeming to boil up famously; but does it boil? Let us see whether that which goes out is steam or not. I think you will soon see the jar (F) will be filled with vapour, if that which rises from the water is steam. But can it be steam? Why, certainly not; because there it remains, you see, unchanged. There it is standing over the water, and it cannot therefore be steam, but must be a permanent gas of some sort. What is it? Is it hydrogen? Is it anything else? Well, we will examine it. If it is hydrogen, it will burn. [*The Lecturer then ignited a portion of the gas collected, which burnt with an explosion.*] It is certainly something combustible, but

Fig. 19

not combustible in the way that hydrogen is. Hydrogen would not have given you that noise; but the colour of

that light, when the thing did burn, was like that of hydrogen : it will, however, burn without contact with the air. That is why I have chosen this other form of apparatus, for the purpose of pointing out to you what are the particular circumstances of this experiment. In place of an open vessel I have taken one that is closed (our battery is so beautifully active that we are even boiling the mercury, and getting all things right—not wrong, but vigorously right) ; and I am going to shew you that that gas, whatever it may be, can burn without air, and in that respect differs from a candle, which cannot burn without the air. And our manner of doing this is as follows :—I have here a glass vessel (G) which is fitted with two platinum-wires (I K), through which I can apply electricity ; and we can put the vessel on the air-pump and exhaust the air, and when we have taken the air out we can bring it here and fasten it on to this jar (F), and let into the vessel that gas which was formed by the action of the voltaic battery upon the water, and which we have produced by changing the water into it,—for I may go as far as this, and say we have really, by that experiment, changed the water into that gas. We have not only altered its condition, but we have changed it really and truly into that gaseous substance, and all the water is there which was decomposed by the experiment. As I screw this vessel (G H) on here (H), and make the tubes well connected, and when I open the stopcocks (H H H), if you watch the level of the water (in F), you will see that the gas will rise. I will now close the stopcocks, as I have drawn up as much as the vessel can hold, and being safely conveyed into that chamber, I will pass into it an electric spark from this Leyden jar (L), when the vessel, which is now quite clear and bright, will become dim. There will be no sound, for the vessel is strong enough to confine the explosion. [*A spark was then passed through the jar, when the explosive mixture was ignited.*] Did you see that brilliant light? If I again screw the vessel on to the jar, and open these stopcocks,

you will see that the gas will rise a second time. [*The stopcocks were then opened.*] Those gases [*referring to the gases first collected in the jar, and which had just been ignited by the electric spark*] have disappeared, as you see : their place is vacant, and fresh gas has gone in. Water has been formed from them ; and if we repeat our operation [*repeating the last experiment*], I shall have another vacancy, as you will see by the water rising. I always have an empty vessel after the explosion, because the vapour or gas into which that water has been resolved by the battery explodes under the influence of the spark, and changes into water ; and by-and-by you will see in this upper vessel some drops of water trickling down the sides and collecting at the bottom.

We are here dealing with water entirely, without reference to the atmosphere. The water of the candle had the atmosphere helping to produce it : but in this way it can be produced independently of the air. Water, therefore, ought to contain that other substance which the candle takes from the air, and which, combining with the hydrogen, produces water.

Fig. 20

Just now you saw that one end of this battery took hold of the copper, extracting it from the vessel which contained the blue solution. It was effected by this wire ;

and surely we may say, if the battery has such power with a metallic solution which we made and unmade, may we not find that it is possible to split asunder the component parts of the water, and put them into this place and that place? Suppose I take the poles—the metallic ends of this battery—and see what will happen with the water in this apparatus (*fig.* 20), where we have separated the two ends far apart. I place one here (at A), and the other there (at B), and I have little shelves with holes which I can put upon each pole, and so arrange them that whatever escapes from the two ends of the battery will appear as separate gases; for you saw that the water did not become vaporous, but gaseous. The wires are now in perfect and proper connection with the vessel containing the water; and you see the bubbles rising: let us collect these bubbles and see what they are. Here is a glass cylinder (o); I fill it with water and put it over one end (A) of the pile; and I will take another (H) and put it over the other end (B) of the pile. And so now we have a double apparatus, with both places delivering gas. Both these jars will fill with gas. There they go, that to the right (H) filling very rapidly; the one to the left (o) filling not so rapidly; and though I have allowed some bubbles to escape, yet still the action is going on pretty regularly; and were it not that one is rather smaller than the other, you would see that I should have twice as much in this (H) as I have in that (o). Both these gases are colourless; they stand over the water without condensing; they are alike in all things—I mean in all *apparent* things; and we have here an opportunity of examining these bodies and ascertaining what they are. Their bulk is large, and we can easily apply experiments to them. I will take this jar (H) first, and will ask you to be prepared to recognise hydrogen.

Think of all its qualities—the light gas which stood well in inverted vessels, burning with a pale flame at the mouth of the jar—and see whether this gas does not satisfy

all these conditions. If it be hydrogen, it will remain here while I hold this jar inverted. [*A light was then applied, when the hydrogen burnt.*] What is there now in the other jar? You know that the two together made an explosive mixture. But what can this be which we find as the other constituent in water, and which must therefore be that substance which made the hydrogen burn? We know that the water we put into the vessel consisted of the two things together. We find one of these in hydrogen : what must that other be which was in the water before the experiment, and which we now have by itself? I am about to put this lighted splinter of wood into the gas. The gas itself will not burn, but it will make the splinter of wood burn. [*The Lecturer ignited the end of the wood, and introduced it into the jar of gas.*] See how it invigorates the combustion of the wood, and how it makes it burn far better than the air would make it burn ; and now you see by itself that every other substance which is contained in the water, and which, when the water was formed by the burning of the candle, must have been taken from the atmosphere. What shall we call it, A, B, or C? Let us call it O—call it "Oxygen": it is a very good distinct-sounding name. This, then, is the oxygen which was present in the water, forming so large a part of it.

We shall now begin to understand more clearly our experiments and researches ; because, when we have examined these things once or twice, we shall soon see why a candle burns in the air. When we have in this way analysed the water—that is to say, separated, or electrolysed its parts out of it—we get two volumes of hydrogen, and one of the body that burns it. And these two are represented to us on the following diagram, with their weights also stated ; and we shall find that the oxygen is a very heavy body by comparison with the hydrogen. It is the other element in water.

I had better, perhaps, tell you now how we get this oxygen abundantly, having shewn you how we can separate it from

the water. Oxygen, as you will immediately imagine, exists in the atmosphere ; for how should the candle burn to produce water without it ? Such a thing would be absolutely

1 Hydrogen	8 Oxygen
	9

Oxygen	88·9
Hydrogen	.	.	.		11·1
Water	100·0

impossible, and chemically impossible, without oxygen. Can we get it from the air ? Well, there are some very complicated and difficult processes by which we can get it from the air ; but we have better processes. There is a substance

Fig. 21

called the black oxide of manganese : it is a very black-looking mineral, but very useful, and when made red-hot it gives out oxygen. Here is an iron bottle which has had some of this substance put into it, and there is a tube

fixed to it, and a fire ready made, and Mr. Anderson will put that retort into the fire, for it is made of iron, and can stand the heat. Here is a salt called chlorate of potassa, which is now made in large quantities for bleaching, and chemical and medical uses, and for pyrotechnic and other purposes. I will take some and mix it with some of the oxide of manganese (oxide of copper, or oxide of iron would do as well); and if I put these together in a retort, far less than a red heat is sufficient to evolve this oxygen from the mixture. I am not preparing to make much, because we only want sufficient for our experiments; only, as you will see immediately, if I use too small a charge, the first portion of the gas will be mixed with the air already in the retort, and I should be obliged to sacrifice the first portion of the gas, because it would be so much diluted with air; the first portion must therefore be thrown away. You will find in this case, that a common spirit-lamp is quite sufficient for me to get the oxygen, and so we shall have two processes going on for its preparation. See how freely the gas is coming over from that small portion of the mixture. We will examine it, and see what are its properties. Now, in this way we are producing, as you will observe, a gas just like the one we had in the experiment with the battery, transparent, undissolved by water, and presenting the ordinary visible properties of the atmosphere. (As this first jar contains the air, together with the first portions of the oxygen set free during the preparation, we will carry it out of the way, and be prepared to make our experiments in a regular, dignified manner.) And, inasmuch as that power of making wood, wax, or other things burn, was so marked in the oxygen we obtained by means of the voltaic battery from water, we may expect to find the same property here. We will try it. You see there is the combustion of a lighted taper in air, and here is its combustion in this gas [*lowering the taper into the jar*]. See how brightly and how beautifully it burns! You can also see more than this,—you

70

will perceive it is a heavy gas, whilst the hydrogen would go up like a balloon, or even faster than a balloon, when not encumbered with the weight of the envelope. You may easily see that although we obtained from water

Fig. 22

twice as much in volume of the hydrogen as of oxygen, it does not follow that we have twice as much in weight— because one is heavy, and the other a very light gas. We have means of weighing gases or air ; but without stopping to explain that, let me just tell you what their respective weights are. The weight of a pint of hydrogen is three-quarters of a grain ; the weight of the same quantity of oxygen is nearly twelve grains. This is a very great difference. The weight of a cubit foot of hydrogen is one-twelfth of an ounce ; and the weight of a cubit foot of oxygen is one ounce and a third. And so on we might come to masses of matter which may be weighed in the balance, and which we can take account of as to hundred-weights and as to tons, as you will see almost immediately.

Now, as regards this very property of oxygen supporting combustion, which we may compare to air, I will take a

piece of candle to shew it you in a rough way, and the result will be rough. There is our candle burning in the air: how will it burn in oxygen? I have here a jar of this gas, and I am about to put it over the candle for you to compare the action of this gas with that of the air. Why, look at it: it looks something like the light you saw at the poles of the voltaic battery. Think how vigorous that action must be! And yet, during all that action, nothing more is produced than what is produced by the burning of the candle in air. We have the same production of water, and the same phenomena exactly, when we use this gas instead of air, as we have when the candle is burnt in air.

But now we have got a knowledge of this new substance, we can look at it a little more distinctly, in order to satisfy ourselves that we have got a good general understanding of this part of the product of a candle. It is wonderful how great the supporting powers of this substance are as regards combustion. For instance, here is a lamp which, simple though it be, is the original, I may say, of a great variety of lamps which are constructed for divers purposes —for light-houses, microscopic illuminations, and other uses; and if it were proposed to make it burn very brightly, you would say, "If a candle burnt better in oxygen, will not a lamp do the same?" Why, it will do so. Mr. Anderson will give me a tube coming from our oxygen reservoir, and I am about to apply it to this flame, which I will previously make burn badly on purpose. There comes the oxygen: what a combustion that makes! But if I shut it off, what becomes of the lamp? [*The flow of oxygen was stopped, and the lamp relapsed to its former dimness.*] It is wonderful how, by means of oxygen, we get combustion accelerated. But it does not affect merely the combustion of hydrogen, or carbon, or the candle; but it exalts all combustions of the common kind. We will take one which relates to iron, for instance, as you have already seen iron burn a little in the atmosphere. Here is

IRON WIRE BURNED IN OXYGEN

a jar of oxygen, and this is a piece of iron wire ; but if it were a bar as thick as my wrist, it would burn the same. I first attach a little piece of wood to the iron, I then set the wood on fire, and let them both down together into

Fig. 23

the jar. The wood is now alight, and there it burns as wood should burn in oxygen; but it will soon communicate its combustion to the iron. The iron is now burning brilliantly, and will continue so for a long time. As long as we supply oxygen, so long can we carry on the combustion of the iron, until the latter is consumed.

We will now put that on one side, and take some other substance ; but we must limit our experiments, for we have not time to spare for all the illustrations you would have a right to if we had more time. We will take a piece of sulphur—you know how sulphur burns in the air—well, we put it into the oxygen, and you will see that whatever can burn in air, can burn with a far greater intensity in oxygen, leading you to think that perhaps the atmosphere itself owes all its power of combustion to this gas. The sulphur is now burning very quietly in the oxygen ; but you cannot for a moment mistake the very high and increased action which takes place when it is so burnt, instead of being burnt merely in common air.

I am now about to shew you the combustion of another substance—phosphorus. I can do it better for you here than you can do it at home. This is a very combustible

73

substance ; and if it be so combustible in air, what might you expect it would be in oxygen? I am about to shew it to you not in its fullest intensity, for if I did so we should almost blow the apparatus up—I may even now crack the jar, though I do not want to break things carelessly.

Fig. 24

You see how it burns in the air. But what a glorious light it gives out when I introduce it into oxygen. [*Introducing the lighted phosphorus into the jar of oxygen.*] There you see the solid particles going off, which cause that combustion to be so brilliantly luminous.

Thus far we have tested this power of oxygen, and the high combustion it produces by means of other substances. We must now, for a little while longer, look at it as respects the hydrogen. You know, when we allowed the oxygen and the hydrogen derived from the water to mix and burn together, we had a little explosion. You remember, also, that when I burnt the oxygen and the hydrogen in a jet together, we got very little light, but great heat. I am now about to set fire to oxygen and hydrogen, mixed in the proportion in which they occur in water. Here is a vessel containing one volume of oxygen and two volumes of hydrogen. This mixture is exactly of the same nature as the gas we just now obtained from the voltaic battery : it would be far too much to burn at once ; I have therefore arranged to blow soap bubbles with it, and burn those

bubbles, that we may see by a general experiment or two how this oxygen supports the combustion of the hydrogen. First of all, we will see whether we can blow a bubble. Well, there goes the gas [*causing it to issue through a tobacco-pipe into some soapsuds.*] Here I have a bubble. I am receiving them on my hand : and you will perhaps think I am acting oddly in this experiment; but it is to shew you that we must not always trust to noise and sounds, but rather to real facts. [*Exploding a bubble on the palm of his hand.*] I am afraid to fire a bubble from the end of the pipe, because the explosion would pass up into the jar and blow it to pieces. This oxygen then will unite with the hydrogen, as you see by the phenomena, and hear by the sound, with the utmost readiness of action, and all its powers are then taken up in its neutralisation of the qualities of the hydrogen.

So now I think you will perceive the whole history of water with reference to oxygen and the air, from what we have before said. Why does a piece of potassium decompose water? Because it finds oxygen in the water. What is set free when I put it in the water, as I am about to do again? It sets free hydrogen, and the hydrogen burns ; but the potassium itself combines with oxygen ; and this piece of potassium, in taking the water apart—the water, you may say, derived from the combustion of the candle— takes away the oxygen which the candle took from the air, and so sets the hydrogen free ; and even if I take a piece of ice, and put a piece of potassium upon it, the beautiful affinities by which the oxygen and the hydrogen are related are such, that the ice will absolutely set fire to the potassium. I shew this to you to-day, in order to enlarge your ideas of these things, and that you may see how greatly results are modified by circumstances. There is the potassium on the ice, producing a sort of volcanic action.

It will be my place, when next we meet, having pointed out these anomalous actions, to show you that none of these extra and strange effects are met with by us—that

none of these strange and injurious actions take place when we are burning, not merely a candle, but gas in our streets, or fuel in our fireplaces, so long as we confine ourselves within the laws that Nature has made for our guidance.

LECTURE V

OXYGEN PRESENT IN THE AIR—NATURE OF THE ATMO-
SPHERE—ITS PROPERTIES—OTHER PRODUCTS FROM THE
CANDLE—CARBONIC ACID—ITS PROPERTIES

WE have now seen that we can produce hydrogen and
oxygen from the water that we obtained from the candle.
Hydrogen, you know, comes from the candle, and oxygen,
you believe, comes from the air. But then you have a
right to ask me, "How is it that the air and the oxygen
do not equally well burn the candle?" If you remember
what happened when I put a jar of oxygen over a piece
of candle, you recollect there was a very different kind of
combustion to that which took place in the air. Now, why
is this? It is a very important question, and one I shall
endeavour to make you understand : it relates most inti-
mately to the nature of the atmosphere, and is most important
to us.

We have several tests for oxygen besides the mere
burning of bodies. You have seen a candle burnt in
oxygen, or in the air ; you have seen phosphorus burnt in
the air, or in oxygen ; and you have seen iron-filings burnt
in oxygen. But we have other tests besides these, and I
am about to refer to one or two of them for the purpose
of carrying your conviction and your experience further.
Here we have a vessel of oxygen. I will shew its presence
to you : if I take a little spark and put it into that oxygen,
you know, by the experience you gained the last time we
met, what will happen ; if I put that spark into the jar,
it will tell you whether we have oxygen here or not. Yes !

77

We have proved it by combustion; and now here is another test for oxygen, which is a very curious and useful one. I have here two jars full of gas, with a plate between them to prevent their mixing; I take the plate away, and the gases are creeping one into the other. "What happens?" say you: "they together produce no such combustion as was seen in the case of the candle." But see how the presence of oxygen is told by its association with this other substance. What a beautifully coloured gas I have obtained in this way, shewing me the presence of the oxygen! In the same way we can try this experiment by mixing common air with this test gas. Here is a jar containing air—such air as the candle would burn in—and here is a jar or bottle containing the test gas. I let them come together over water, and you see the result: the contents of the test bottle are flowing into the jar of air, and you see I obtain exactly the same kind of action as before, and that shews me that there is oxygen in the air—the very same substance that has been already obtained by us from the water produced by the candle. But then, beyond that, how is it that the candle does not burn in air as well as in oxygen? We will come to that point at once. I have here two jars; they are filled to the same height with gas, and the appearance to the eye is alike in both, and I really do not know at present which of these jars contains oxygen and which contains air, although I know they have previously been filled with these gases. But here is our test gas, and I am going to work with the two jars, in order to examine whether there is any difference between them in the quality of reddening this gas. I am now going to turn this test gas into one of the jars, and observe what happens. There is reddening, you see; there is then oxygen present. We will now test the other jar; but you see this is not so distinctly red as the first: and, further, this curious thing happens,—if I take these two gases and shake them well together with water, we shall absorb the red gas; and then, if I put

78

in more of this test gas and shake again, we shall absorb more; and I can go on as long as there be any oxygen present to produce that effect. If I let in air, it will not matter; but the moment I introduce water, the red gas disappears; and I may go on in this way, putting in more and more of the test gas, until I come to something left behind which will not redden any longer by the use of that particular body that rendered the air and the oxygen red. Why is that? You see in a moment it is because there is, besides oxygen, something else present which is left behind. I will let a little more air into the jar, and if it turns red you will know that some of that reddening gas is still present, and that consequently it was not for the want of this producing body that that air was left behind.

Now, you will begin to understand what I am about to say. You saw that when I burnt phosphorus in a jar, as the smoke produced by the phosphorus and the oxygen of the air condensed, it left a good deal of gas unburnt, just as this red gas left something untouched—there was, in fact, this gas left behind, which the phosphorus cannot touch, which the reddening gas cannot touch, and this something is not oxygen, and yet is part of the atmosphere.

So that is one way of opening out air into the two things of which it is composed—oxygen, which burns our candles, our phosphorus, or anything else; and this other substance —nitrogen—which will not burn them. This other part of the air is by far the larger proportion, and it is a very curious body, when we come to examine it; it is remarkably curious, and yet you say, perhaps, that it is very uninteresting. It is uninteresting in some respects because of this—that it shews no brilliant effects of combustion. If I test it with a taper as I do oxygen and hydrogen, it does not burn like hydrogen, nor does it make the taper burn like oxygen. Try it in any way I will, it does neither the one thing nor the other: it will not take fire; it will not let the taper burn; it puts out the

combustion of everything. There is nothing that will burn in it in common circumstances. It has no smell; it is not sour; it does not dissolve in water; it is neither an acid nor an alkali; it is as indifferent to all our organs as it is possible for a thing to be. And you might say, "It is nothing; it is not worth chemical attention; what does it do in the air?" Ah! then come our beautiful and fine results shewn us by an observant philosophy. Suppose, in place of having nitrogen, or nitrogen and oxygen, we had pure oxygen as our atmosphere; what would become of us? You know very well that a piece of iron lit in a jar of oxygen goes on burning to the end. When you see a fire in an iron grate, imagine where the grate would go to if the whole of the atmosphere were oxygen. The grate would burn up more powerfully than the coals—for the iron of the grate itself is even more combustible than the coals which we burn in it. A fire put into the middle of a locomotive would be a fire in a magazine of fuel, if the atmosphere were oxygen. The nitrogen lowers it down and makes it moderate and useful for us, and then, with all that, it takes away with it the fumes that you have seen produced from the candle, disperses them throughout the whole of the atmosphere, and carries them away to places where they are wanted to perform a great and glorious purpose of good to man, for the sustenance of vegetation; and thus does a most wonderful work, although you say, on examining it, "Why, it is a perfectly indifferent thing." This nitrogen in its ordinary state is an inactive element; no action short of the most intense electric force, and then in the most infinitely small degree, can cause the nitrogen to combine directly with the other element of the atmosphere, or with other things round about it; it is a perfectly indifferent, and therefore, so to say, a safe substance.

But before I take you to that result, I must tell you about the atmosphere itself. I have written on this diagram the composition of one hundred parts of atmospheric air:—

COMPOSITION OF AIR

							Bulk	Weight
Oxygen	20	22.3
Nitrogen	80	77.7
							100	100.0

It is a true analysis of the atmosphere, so far as regards the quantity of oxygen and the quantity of nitrogen present. By our analysis, we find that 5 pints of the atmosphere contain only 1 pint of oxygen, and 4 pints, or 4 parts, of nitrogen by bulk. That is our analysis of the atmosphere. It requires all that quantity of nitrogen to reduce the oxygen down, so as to be able to supply the candle properly with fuel, so as to supply us with an

Fig. 25

atmosphere which our lungs can healthily and safely breathe; for it is just as important to make the oxygen

right for us to breathe, as it is to make the atmosphere right for the burning of the fire and the candle.

But now for this atmosphere. First of all, let me tell you the weight of these gases. A pint of nitrogen weighs $10\frac{4}{10}$ grains, or a cubic foot weighs $1\frac{1}{8}$ ounce. That is the weight of the nitrogen. The oxygen is heavier: a pint of it weighs $11\frac{9}{10}$ grains, and a cubic foot weighs $1\frac{1}{4}$ ounce. A pint of air weighs about $10\frac{7}{10}$ grains, and a cubic foot $1\frac{1}{8}$ ounce.

You have asked me several times, and I am very glad you have, "How do you weigh gases?" I will shew you; it is very simple, and easily done. Here is a balance, and here a copper bottle, made as light as we can consistent with due strength, turned very nicely in the lathe, and made perfectly air-tight, with a stopcock, which we can open and shut, which at present is open, and therefore allows the bottle to be full of air. I have here a nicely adjusted balance, in which I think the bottle in its present condition will be balanced by the weight on the other side. And here is a pump by which we can force the air into this bottle, and with it we will force in a certain number of volumes of air, as measured by the pump. [*Twenty measures were pumped in.*] We will shut that in and put it in the balance. See how it sinks: it is much heavier than it was. By what? By the air that we have forced into it by the pump. There is not a greater *bulk* of air, but there is the same bulk of *heavier* air, because we have forced in air upon it. And that you may have a fair notion in your mind as to how much this air measures, here is a jar full of water. We will open that copper vessel into this jar, and let the air return to its former state. All I have to do now is to screw them tightly together, and to turn the taps, when there, you see, is the bulk of the twenty pumps of air which I forced into the bottle; and to make sure that we have been quite correct in what we have been doing, we will take the bottle again to the balance, and, if it is now

counterpoised by the original weight, we shall be quite sure we have made our experiment correctly. It is balanced; so, you see, we can find out the weight of the extra volumes of air forced in, in that way, and by that means we are able

Fig. 26

to ascertain that a cubic foot of air weighs 1¼ ounce. But that small experiment will by no means convey to your mind the whole literal truth of this matter. It is wonderful how it accumulates when you come to larger volumes. This bulk of air [*a cubic foot*] weighs 1¼ ounce. What do you think of the contents of that box above there, which I have had made for the purpose? The air which is within that box weighs one pound—a full pound; and I have calculated the weight of the air in this room,—you would hardly imagine it, but it is above a ton. So rapidly do the weights rise up, and so important is the presence of the atmosphere, and of the oxygen and the nitrogen

in it, and the use it performs in conveying things to and
fro from place to place, and carrying bad vapours to places
where they will do good instead of harm.

Having given you that little illustration with respect
to the weight of the air, let me shew you certain conse-
quences of it. You have a right to them, because you would
not understand so much without it. Do you remember
this kind of experiment? Have you ever seen it? Suppose
I take a pump somewhat similar to the one I had a
little while ago to force air into the bottle, and suppose
I place it in such a manner that by certain arrangements
I can apply my hand to it : my hand moves about in the

Fig. 27

air so easily that it seems to feel nothing, and I can hardly
get velocity enough by any motion of my own in the
atmosphere to make sure that there is much resistance
to it. But, when I put my hand here [*on the air-pump
receiver, which was afterwards exhausted*], you see what
happens. Why is my hand fastened to this place, and
why am I able to pull this pump about? And see! how
is it that I can hardly get my hand away? Why is this?
It is the weight of the air—the weight of the air that
is above. I have another experiment here, which I think
will explain to you more about it. When the air is
pumped from underneath the bladder which is stretched
over this glass, you will see the effect in another shape :
the top is quite flat at present, but I will make a very
little motion with the pump, and now look at it—see

how it has gone down, see how it is bent in. You will see the bladder go in more and more, until at last I expect it will be driven in and broken by the force of the atmosphere pressing upon it. [*The bladder at last broke with a loud report.*] Now, that was done entirely by the weight of the air pressing on it, and you can easily understand how that is. The particles that are piled up in the atmosphere stand upon each other, as these five cubes do. You can easily conceive that four of these five cubes are resting upon the bottom one, and if I take that away,

Fig. 28

the others will all sink down. So it is with the atmosphere : the air that is above is sustained by the air that is beneath ; and when the air is pumped away from beneath them, the change occurs which you saw when I placed my hand on the air-pump, and which you saw in the case of the bladder, and which you shall see better here. I have tied over this jar a piece of sheet india-rubber, and I am now about to take away the air from the inside of the jar ; and if you will watch the india-rubber—which acts as a partition between the air below and the air above—you will see, when I pump, how the pressure shews itself. See where

it is going to—I can actually put my hand into the jar; and yet this result is only caused by the great and powerful action of the air above. How beautifully it shews this curious circumstance!

Here is something that you can have a pull at, when I have finished to-day. It is a little apparatus of two hollow brass hemispheres, closely fitted together, and having connected with it a pipe and a cock, through which we can exhaust the air from the inside; and although the two halves are so easily taken apart, while the air is left within, yet you will see, when we exhaust it by-and-by, no power of any two of you will be able to pull them apart. Every square inch of surface that is contained in the area of that vessel sustains fifteen pounds by weight, or nearly so, when the air is taken out; and you may try your strength presently in seeing whether you can overcome that pressure of the atmosphere.

Here is another very pretty thing—the boys' sucker, only refined by the philosopher. We young ones have a perfect right to take toys, and make them into philosophy, inasmuch as now-a-days we are turning philosophy into toys. Here is a sucker, only it is made of india-rubber: if I clap it upon the table, you see at once it holds. Why does it hold? I can slip it about, and yet if I try to pull it up, it seems as if it would pull the table with it. I can easily make it slip about from place to place; but only when I bring it to the edge of the table can I get it off. It is only kept down by the pressure of the atmosphere above. We have a couple of them; and if you take these two and press them together, you will see how firmly they stick. And, indeed, we may use them as they are proposed to be used, to stick against windows, or against walls. where they will adhere for an evening, and serve to hang anything on that you want. I think, however, that you boys ought to be shewn experiments that you can make at home; and so here is a very pretty experiment in illustration of the pressure of the atmosphere. Here is a tumbler

of water. Suppose I were to ask you to turn that tumbler upside-down, so that the water shall not fall out, and yet not be kept in by your hand, but merely by using the pressure of the atmosphere. Could you do that? Take a wine-glass, either quite full or half-full of water, and put a flat card on the top, turn it upside down, and then see what becomes of the card and of the water. The air cannot get in because the water by its capillary attraction round the edge keeps it out.

I think this will give you a correct notion of what you may call the materiality of the air; and when I tell you that the box holds a pound of it, and this room more than a ton, you will begin to think that air is something very serious. I will make another experiment, to convince you of this positive resistance. There is that beautiful experiment of the popgun, made so well and so easily, you know, out of a quill, or a tube, or anything of that kind,— where we take a slice of potato, for instance, or an apple, and take the tube and cut out a pellet, as I have now done, and push it to one end. I have made that end tight; and now I take another piece and put it in : it will confine the air that is within the tube perfectly and completely for our purpose ; and I shall now find it absolutely impossible by any force of mine to drive that little pellet close up to the other. It cannot be done. I may press the air to a certain extent, but if I go on pressing, long before it comes to the second, the confined air will drive the front one out with a force something like that of gunpowder ; for gunpowder is in part dependent upon the same action that you see here exemplified.

I saw the other day an experiment which pleased me much, as I thought it would serve our purpose here. (I ought to have held my tongue for four or five minutes before beginning this experiment, because it depends upon my lungs for success.) By the proper application of air I expect to be able to drive this egg out of one cup into the other by the force of my breath ; but if I fail, it is in

a good cause; and I do not promise success, because I have been talking more than I ought to do to make the experiment succeed. [*The Lecturer here tried the experiment, and succeeded in blowing the egg from one egg-cup to the other.*]

You see that the air which I blow goes downwards between the egg and the cup, and makes a blast under the egg, and is thus able to lift a heavy thing—for a full egg is a very heavy thing for air to lift. If you want to make the experiment, you had better boil the egg quite hard first, and then you may very safely try to blow it from one cup to the other, with a little care.

I have now kept you long enough upon this property of the weight of the air, but there is another thing I should like to mention. You saw the way in which, in this popgun, I was able to drive the second piece of potato half or two-thirds of an inch before the first piece started, by virtue of the elasticity of the air—just as I pressed into the copper bottle the particles of air by means of the pump. Now, this depends upon a wonderful property in the air, namely, its elasticity; and I should like to give you a good illustration of this. If I take anything that confines the air properly, as this membrane, which also is able to contract and expand so as to give us a measure of the elasticity of the air, and confine in this bladder a certain portion of air; and then, if we take the atmosphere off from the outside of it, just as in these cases we put the pressure on—if we take the pressure off, you will see how it will then go on expanding and expanding, larger and larger, until it will fill the whole of this bell-jar, shewing you that wonderful property of the air, its elasticity, its compressibility, and expansibility, to an exceedingly large extent, and which is very essential for the purposes and services it performs in the economy of creation.

We will now turn to another very important part of our subject, remembering that we have examined the

candle in its burning, and have found that it gives rise to various products. We have the products, you know, of soot, of water, and of something else which you have not yet examined. We have collected the water, but have allowed the other things to go into the air. Let us now examine some of these other products.

Here is an experiment which I think will help you in part in this way. We will put our candle there, and place over it a chimney, thus. I think my candle will go on

Fig. 29

burning, because the air passage is open at the bottom and the top. In the first place, you see the moisture appearing —that you know about. It is water produced from the candle by the action of the air upon its hydrogen. But, besides that, something is going out at the top: it is not moisture—it is not water—it is not condensable; and yet, after all, it has very singular properties. You will find that the air coming out of the top of our chimney is nearly sufficient to blow out the light I am holding to it; and if I put the light fairly opposed to the current, it will blow

it quite out. You will say that is as it should be ; and I am supposing that you think it ought to do so, because the nitrogen does not support combustion, and ought to put the candle out, since the candle will not burn in nitrogen. But is there nothing else there than nitrogen ? I must now anticipate—that is to say, I must use my own knowledge to supply you with the means that we adopt for the purpose of ascertaining these things, and examining such gases as these. I will take an empty bottle —here is one—and if I hold it over this chimney, I shall get the combustion of the candle below sending its results into the bottle above ; and we shall soon find that this bottle contains, not merely an air that is bad as regards the combustion of a taper put into it, but having other properties.

Let me take a little quicklime and pour some common water on to it—the commonest water will do. I will stir it a moment, then pour it upon a piece of filtering paper in a funnel, and we shall very quickly have a clear water proceeding to the bottle below, as I have here. I have plenty of this water in another bottle ; but, nevertheless, I should like to use the lime-water that was prepared before you, so that you may see what its uses are. If I take some of this beautiful clear lime-water, and pour it into this jar, which has collected the air from the candle, you will see a change coming about. Do you see that the water has become quite milky ? Observe, that will not happen with air merely. Here is a bottle filled with air ; and if I put a little lime-water into it, neither the oxygen nor the nitrogen, nor anything else that is in that quantity of air, will make any change in the lime-water. It remains perfectly clear, and no shaking of that quantity of lime-water with that quantity of air in its common state will cause any change ; but if I take this bottle with the lime-water, and hold it so as to get the general products of the candle in contact with it, in a very short time we shall have it milky. There is the chalk, consisting of the

lime which we used in making the lime-water, combined with something that came from the candle—that other product which we are in search of, and which I want to tell you about to-day. This is a substance made visible to us by its action, which is not the action of the lime-water either upon the oxygen or upon the nitrogen, nor upon the water itself, but it is something new to us from the candle. And then we find this white powder, produced by the lime-water and the vapour from the candle, appears to us very much like whitening or chalk, and, when examined, it does prove to be exactly the same substance as whitening or chalk. So we are led, or have been led, to observe upon the various circumstances of this experiment, and to trace this production of chalk to its various causes, to give us the true knowledge of the nature of this combustion of the candle—to find that this substance, issuing from the candle, is exactly the same as that substance which would issue from a retort, if I were to put some chalk into it with a little moisture, and make it red-hot : you would then find that exactly the same substance would issue from it as from the candle.

But we have a better means of getting this substance, and in greater quantity, so as to ascertain what its general characters are. We find this substance in very great abundance in a multitude of cases where you would least expect it. All limestones contain a great deal of this gas which issues from the candle, and which we call *carbonic acid.* All chalks, all shells, all corals contain a great quantity of this curious air. We find it fixed in these stones ; for which reason Dr. Black called it " fixed air " —finding it in these fixed things like marble and chalk. He called it fixed air, because it lost its quality of air, and assumed the condition of a solid body. We can easily get this air from marble. Here is a jar containing a little muriatic acid, and here is a taper which, if I put it into that jar, will shew only the presence of common air. There is, you see, pure air down to the bottom ; the jar is full

of it. Here is a substance—marble, a very beautiful and superior marble—and if I put these pieces of marble into the jar, a great boiling apparently goes on. That, however, is not steam—it is a gas that is rising up; and if I now search the jar by a candle, I shall have exactly the same effect produced upon the taper as I had from the air which issued from the end of the chimney over the burning candle. It is exactly the same action, and caused by the very same substance that issued from the candle; and in this way we can get carbonic acid in great abundance—we have already nearly filled the jar. We also find that this gas is not merely contained in marble. Here is a vessel in which I have put some common whitening—chalk, which has been washed in water and deprived of its coarser particles, and so supplied to the plasterer as whitening. Here is a large jar containing this whitening and water, and I have here some strong sulphuric acid, which is the acid you might have to use if you were to make these experiments (only, in using this acid with limestone, the body that is produced is an insoluble substance, whereas the muriatic acid produces a soluble substance that does not so much thicken the water). And you will seek out a reason why I take this kind of apparatus for the purpose of shewing this experiment. I do it because you may repeat in a small way what I am about to do in a large one. You will have here just the same kind of action; and I am evolving in this large jar carbonic acid, exactly the same in its nature and properties as the gas which we obtained from the combustion of the candle in the atmosphere. And no matter how different the two methods by which we prepare this carbonic acid, you will see, when we get to the end of our subject, that it is all exactly the same, whether prepared in the one way or in the other.

We will now proceed to the next experiment with regard to this gas. What is its nature? Here is one of the vessels full, and we will try it, as we have done so many other gases, by combustion. You see it is not

combustible, nor does it support combustion. Neither, as we know, does it dissolve much in water, because we collect it over water very easily. Then, you know that it has an effect, and becomes white in contact with lime-water; and when it does become white in that way, it becomes one of the constituents to make carbonate of lime or limestone.

The next thing I must shew you is, that it really does dissolve a little in water, and therefore that it is unlike oxygen and hydrogen in that respect. I have here an apparatus by which we can produce this solution. In the lower part of this apparatus is marble and acid, and in the upper part cold water. The valves are so arranged that the gas can get from one to the other. I will set it in action now, and you can see the gas bubbling up through the water, as it has been doing all night long, and by this time we shall find that we have this substance dissolved in the water. If I take a glass and draw off some of the water, I find that it tastes a little acid to the mouth: it is impregnated with carbonic acid; and if I now apply a little lime-water to it, that will give us a test of its presence. This water will make the lime-water turbid and white, which is proof of the presence of carbonic acid.

Then it is a very weighty gas—it is heavier than the atmosphere. I have put their respective weights at the lower part of this table, along with, for comparison, the weights of the other gases we have been examining :—

	Pint	Cubic Foot
Hydrogen . . .	$\frac{3}{4}$ grains	$\frac{1}{12}$ ounce
Oxygen	$11\frac{9}{10}$,,	$1\frac{1}{3}$,,
Nitrogen	$10\frac{4}{10}$,,	$1\frac{1}{6}$,,
Air	$10\frac{7}{10}$,,	$1\frac{1}{5}$,,
Carbonic acid . .	$16\frac{1}{3}$,,	$1\frac{9}{10}$,,

A pint of it weighs $16\frac{1}{3}$ grains, and a cubic foot weighs $1\frac{9}{10}$ ounce, almost two ounces. You can see by many experiments that this is a heavy gas. Suppose I take a glass

containing nothing else but air, and from this vessel containing the carbonic acid I attempt to pour a little of this gas into that glass; I wonder whether any has gone in or not. I cannot tell by the appearance, but I can in this way [*introducing the taper*]. Yes, there it is, you see; and if I were to examine it by lime-water, I should find it by that test also. I will take this little bucket, and

Fig. 30

put it down into the well of carbonic acid—indeed, we too often have real wells of carbonic acid—and now, if there is any carbonic acid, I must have got to it by this time, and it will be in this bucket, which we will examine with a taper. There it is, you see; it is full of carbonic acid.

There is another experiment by which I will shew you its weight. I have here a jar suspended at one end of a balance—it is now equipoised; but when I pour this carbonic acid into the jar on the one side which now contains air, you will see it sink down at once, because of the carbonic acid that I pour into it. And now, if I examine this jar with the lighted taper, I shall find that the carbonic acid has fallen into it, and it no longer has any power of

supporting the combustion. If I blow a soap-bubble, which of course will be filled with air, and let it fall into this jar of carbonic acid, it will float. But I shall first of all take one of these little balloons filled with air. I am not quite sure where the carbonic acid is ; we will just try the depth, and see whereabouts is its level. There, you see, we have this bladder floating on the carbonic acid ; and if I evolve some more of the carbonic acid, the bladder

Fig. 31

will be lifted up higher. There it goes—the jar is nearly full ; and now I will see whether I can blow a soap-bubble on that, and float it in the same way. [*The Lecturer here blew a soap-bubble, and allowed it to fall into the jar of carbonic acid, when it floated in it midway.*] It is floating, as the balloon floated, by virtue of the greater weight of the carbonic acid than of the air. And now, having so far given you the history of the carbonic acid—as to its sources in the candle, as to its physical properties and weight— when we next meet I shall shew you of what it is composed, and where it gets its elements from.

LECTURE VI

CARBON OR CHARCOAL—COAL-GAS—RESPIRATION AND ITS
ANALOGY TO THE BURNING OF A CANDLE—CONCLU-
SION

A LADY, who honours me by her presence at these Lectures, has conferred a still further obligation by sending me these two candles, which are from Japan, and, I presume, are made of that substance to which I referred in a former lecture. You see that they are even far more highly ornamented than the French candles; and, I suppose, are candles of luxury, judging from their appearance. They have a remarkable peculiarity about them,—namely, a hollow wick,—that beautiful peculiarity which Argand introduced into the lamp, and made so valuable. To those who receive such presents from the East, I may just say that this and such like materials gradually undergo a change which gives them on the surface a dull and dead appearance; but they may easily be restored to their original beauty, if the surface be rubbed with a clean cloth or silk handkerchief, so as to polish the little rugosity or roughness: this will restore the beauty of the colours. I have so rubbed one of these candles, and you see the difference between it and the other which has not been polished, but which may be restored by the same process. Observe, also, that these moulded candles from Japan are made more conical than the moulded candles in this part of the world.

I told you, when we last met, a good deal about carbonic acid. We found, by the lime-water test, that when the

vapour from the top of the candle or lamp was received into bottles, and tested by this solution of lime-water (the composition of which I explained to you, and which you can make for yourselves), we had that white opacity which was in fact calcareous matter, like shells and corals, and many of the rocks and minerals in the earth. But I have not yet told you fully and clearly the chemical history of this substance—carbonic acid—as we have it from the candle, and I must now resume that subject. We have seen the products, and the nature of them, as they issue from the candle. We have traced the water to its elements, and now we have to see where are the elements of the carbonic acid supplied by the candle. A few experiments will shew this. You remember that, when a candle burns badly, it produces smoke; but if it is burning well, there is no smoke. And you know that the brightness of the candle is due to this smoke, which becomes ignited. Here is an experiment to prove this: so long as the smoke remains in the flame of the candle and becomes ignited, it gives a beautiful light, and never appears to us in the form of black particles. I will light some fuel, which is extravagant in its burning. This will serve our purpose—a little turpentine on a sponge. You see the smoke rising from it, and floating into the air in large quantities; and, remember now, the carbonic acid that we have from the candle is from such smoke as that. To make that evident to you, I will introduce this turpentine burning on the sponge into a flask where I have plenty of oxygen, the rich part of the atmosphere, and you now see that the smoke is all consumed. This is the first part of our experiment; and now, what follows? The carbon which you saw flying off from the turpentine flame in the air is now entirely burned in this oxygen, and we shall find that it will, by this rough and temporary experiment, give us exactly the same conclusion and result as we had from the combustion of the candle. The reason why I make the experiment in this manner is solely that I may cause the steps of our

demonstration to be so simple that you can never for a moment lose the train of reasoning, if you only pay attention. All the carbon which is burned in oxygen, or air, comes out as carbonic acid, whilst those particles which are not so burned shew you the second substance in the carbonic acid—namely, the carbon—that body which made the flame so bright whilst there was plenty of air, but which was thrown off in excess when there was not oxygen enough to burn it.

I have also to shew you a little more distinctly the history of carbon and oxygen, in their union to make carbonic acid. You are now better able to understand this than before, and I have prepared three or four experiments by way of illustration. This jar is filled with oxygen, and here is some carbon which has been placed in a crucible, for the purpose of being made red-hot. I keep my jar dry, and venture to give you a result imperfect in some degree, in order that I may make the experiment brighter. I am about to put the oxygen and the carbon together. That this is carbon (common charcoal pulverised), you will see by the way in which it burns in the air [*letting some of the red-hot charcoal fall out of the crucible*]. I am now about to burn it in oxygen gas, and look at the difference. It may appear to you at a distance as if it were burning with a flame; but it is not so. Every little piece of charcoal is burning as a spark, and whilst it so burns it is producing carbonic acid. I specially want these two or three experiments to point out what I shall dwell upon more distinctly by-and-by—that carbon burns in this way, and not as a flame.

Instead of taking many particles of carbon to burn, I will take a rather large piece, which will enable you to see the form and size, and to trace the effects very decidedly. Here is the jar of oxygen, and here is the piece of charcoal, to which I have fastened a little piece of wood, which I can set fire to, and so commence the combustion, which I could not conveniently do without. You now see the

charcoal burning, but not as a flame (or if there be a flame, it is the smallest possible one, which I know the cause of—namely, the formation of a little carbonic oxide close upon the surface of the carbon). It goes on burning, you see, slowly producing carbonic acid by the union of this carbon or charcoal (they are equivalent terms) with the oxygen. I have here another piece of charcoal, a piece of bark, which has the quality of being blown to pieces—exploding as it burns. By the effect of the heat, we shall reduce the lump of carbon into particles that will fly off; still every particle, equally with the whole mass, burns in this peculiar way : it burns as a coal, and not like a flame. You observe a multitude of little combustions going on, but no flame. I do not know a finer experiment than this, to shew that carbon burns with a spark.

Here, then, is carbonic acid formed from its elements. It is produced at once; and if we examined it by lime-water, you will see that we have the same substance which I have previously described to you. By putting together 6 parts of carbon by weight (whether it comes from the flame of a candle or from powdered charcoal) and 16 parts of oxygen by weight, we have 22 parts of carbonic acid; and, as we saw last time, the 22 parts of carbonic acid, combined with 28 parts of lime, produced common carbonate of lime. If you were to examine an oyster-shell, and weigh the component parts, you would find that every 50 parts would give 6 of carbon and 16 of oxygen, combined with 28 of lime. However, I do not want to trouble you with these minutiæ—it is only the general philosophy of the matter that we can now go into. See how finely the carbon is dissolving away [*pointing to the lump of charcoal burning quietly in the jar of oxygen*]. You may say that the charcoal is actually dissolving in the air round about; and if that were perfectly pure charcoal, which we can easily prepare, there would be no residue whatever. When we have a perfectly cleansed and purified piece of carbon, there is no ash left. The carbon burns as a solid dense

body, that heat alone cannot change as to its solidity, and yet it passes away into vapour that never condenses into solid or liquid under ordinary circumstances ; and what is more curious still, is the fact that the oxygen does not change in its bulk by the solution of the carbon in it. Just as the bulk is at first, so it is at last, only it has become carbonic acid.

There is another experiment which I must give you before you are fully acquainted with the general nature of carbonic acid. Being a compound body, consisting of carbon and oxygen, carbonic acid is a body that we ought to be able to take asunder. And so we can. As we did with water, so we can with carbonic acid—take the two parts asunder. The simplest and quickest way is to act upon the carbonic acid by a substance that can attract the oxygen from it, and leave the carbon behind. You recollect that I took potassium and put it upon water or ice, and you saw that it could take the oxygen from the hydrogen. Now, suppose we do something of the same kind here with this carbonic acid. You know carbonic acid to be a heavy gas. I will not test it with lime-water, as that will interfere with our subsequent experiments ; but I think the heaviness of the gas and the power of extinguishing flame will be sufficient for our purpose. I introduce a flame into the gas, and you will see whether it will be put out. You see the light is extinguished. Indeed, the gas may, perhaps, put out phosphorus, which, you know, has a pretty strong combustion. Here is a piece of phosphorus heated to a high degree. I introduce it into gas, and you observe the light is put out ; but it will take fire again in the air, because there it re-enters into combustion. Now, let me take a piece of potassium, a substance which, even at common temperatures, can act upon carbonic acid, though not sufficiently for our present purpose, because it soon gets covered with a protecting coat ; but if we warm it up to the burning point in air, as we have a fair right to do, and as we have done with phosphorus, you will see that

it can burn in carbonic acid ; and if it burns, it will burn by taking oxygen, so that you will see what is left behind. I am going, then, to burn this potassium in the carbonic acid, as a proof of the existence of oxygen in the carbonic acid. [*In the preliminary process of heating, the potassium exploded.*] Sometimes we get an awkward piece of potassium that explodes, or something like it, when it burns. I will take another piece ; and now that it is heated, I introduce it into the jar, and you perceive that it burns in the carbonic acid—not so well as in the air, because the carbonic acid contains the oxygen combined ; but it does burn, and takes away the oxygen. If I now put this potassium into water, I find that, besides the potash formed (which you need not trouble about), there is a quantity of carbon produced. I have here made the experiment in a very rough way ; but I assure you that if I were to make it carefully, devoting a day to it, instead of five minutes, we should get all the proper amount of charcoal left in the spoon, or in the place where the potassium was burned, so that there could be no doubt as to the result. Here, then, is the carbon obtained from the carbonic acid, as a common black substance ; so that you have the entire proof of the nature of carbonic acid as consisting of carbon and oxygen. And now, I may tell you, that *whenever* carbon burns under common circumstances, it produces carbonic acid.

Suppose I take this piece of wood, and put it into a bottle with lime-water. I might shake that lime-water up with wood and the atmosphere as long as I pleased, it would still remain clear as you see it ; but suppose I burn the piece of wood in the air of that bottle. You, of course, know I get water. Do I get carbonic acid ? [*The experiment was performed.*] There it is, you see—that is to say, the carbonate of lime, which results from carbonic acid, and that carbonic acid must be formed from the carbon which comes from the wood, from the candle, or any other thing. Indeed, you have yourselves frequently tried a very pretty

experiment, by which you may see the carbon in wood. If you take a piece of wood, and partly burn it, and then blow it out, you have carbon left. There are things that do not shew carbon in this way. A candle does not shew it, but it contains carbon. Here also is a jar of coal-gas, which produces carbonic acid abundantly. You do not see the carbon, but we can soon shew it to you. I will light it, and as long as there is any gas in this cylinder it will go on burning. You see no carbon, but you see a flame; and because that is bright, it will lead you to guess that there is carbon in the flame. But I will shew it to you by another process. I have some of the same gas in another vessel, mixed with a body that will burn the hydrogen of the gas, but will not burn the carbon. I will light them with a burning taper, and you perceive the hydrogen is consumed, but not the carbon, which is left behind as a dense black smoke. I hope that by these three or four experiments you will learn to see when carbon is present, and understand what are the products of combustion, when gas or other bodies are thoroughly burned in the air.

Before we leave the subject of carbon, let us make a few experiments and remarks upon its wonderful condition as respects ordinary combustion. I have shewn you that the carbon in burning burns only as a solid body, and yet you perceive that, after it is burned, it ceases to be a solid. There are very few fuels that act like this. It is, in fact, only that great source of fuel, the carbonaceous series, the coals, charcoals, and woods, that can do it. I do not know that there is any other elementary substance besides carbon that burns with these conditions; and if it had not been so, what would happen to us? Suppose all fuel had been like iron, which, when it burns, burns into a solid substance. We could not then have such a combustion as you have in this fireplace. Here also is another kind of fuel which burns very well—as well as, if not better, than carbon—so well, indeed, as to take fire of itself when

it is in the air, as you see [*breaking a tube full of lead pyrophorus*]. This substance is lead, and you see how wonderfully combustible it is. It is very much divided, and is like a heap of coals in the fireplace; the air can get to its surface and inside, and so it burns. But why does it not burn in that way now, when it is lying in a mass? [*emptying the contents of the tube in a heap on to a plate of iron*]. Simply because the air cannot get to it. Though it can produce a great heat, the great heat which we want in our furnaces and under our boilers, still that which is produced cannot get away from the portion which remains unburned underneath, and that portion, therefore, is prevented from coming in contact with the atmosphere, and cannot be consumed. How different is that from carbon. Carbon burns just in the same way as this lead does, and so gives an intense fire in the furnace, or wherever you choose to burn it; but then the body produced by its combustion passes away, and the remaining carbon is left clear. I shewed you how carbon went on dissolving in the oxygen, leaving no ash; whereas here [*pointing to the heap of pyrophorus*] we have actually more ash than fuel, for it is heavier by the amount of the oxygen which has united with it. Thus you see the difference betweeen carbon and lead or iron: if we choose iron, which gives so wonderful a result in our application of this fuel, either as light or heat. If, when the carbon burnt, the product went off as a solid body, you would have had the room filled with an opaque substance, as in the case of the phosphorus; but when carbon burns, everything passes up into the atmosphere. It is in a fixed, almost unchangeable condition before the combustion; but afterwards it is in the form of gas, which is very difficult (though we have succeeded) to produce in a solid or a liquid state.

Now, I must take you to a very interesting part of our subject—to the relation between the combustion of a candle and that living kind of combustion which goes on within

us. In every one of us there is a living process of combustion going on very similar to that of a candle; and I must try to make that plain to you. For it is not merely true in a poetical sense—the relation of the life of man to a taper; and if you will follow, I think I can make this clear. In order to make the relation very plain, I have devised a little apparatus which we can soon build up before you. Here is a board and a groove cut in it, and I can close the groove at the top part by a little cover. I can then continue the groove as a channel by a glass tube at each end, there being a free passage through the

Fig. 32

whole. Suppose I take a taper or candle (we can now be liberal in our use of the word "candle," since we understand what it means), and place it in one of the tubes; it will go on, you see, burning very well. You observe that the air which feeds the flame passes down the tube at one end, then goes along the horizontal tube, and ascends the tube at the other end in which the taper is placed. If I stop the aperture through which the air enters, I stop combustion, as you perceive. I stop the supply of air, and consequently the candle goes out. But, now, what will you think of this fact? In a former experiment I shewed you the air going from one burning candle to a second candle. If I took the air proceeding from another

candle, and sent it down by a complicated arrangement into this tube, I should put this burning candle out. But what will you say when I tell you that my breath will put out that candle? I do not mean by blowing at all, but simply that the nature of my breath is such that a candle cannot burn in it. I will now hold my mouth over the aperture, and without blowing the flame in any way, let no air enter the tube but what comes from my mouth. You see the result. I did not blow the candle out. I

Fig. 33

merely let the air which I expired pass into the aperture, and the result was that the light went out for want of oxygen, and for no other reason. Something or other—namely, my lungs—had taken away the oxygen from the air, and there was no more to supply the combustion of the candle. It is, I think, very pretty to see the time it takes before the bad air which I throw into this part of the apparatus has reached the candle. The candle at first goes on burning, but so soon as the air has had time to reach it, it goes out. And, now, I will shew you another

experiment, because this is an important part of our philosophy. Here is a jar which contains fresh air, as you can see by the circumstance of a candle or gaslight burning it. I make it close for a little time, and by means of a pipe I get my mouth over it so that I can inhale the air. By putting it over water, in the way that you see (*fig.* 33), I am able to draw up this air (supposing the cork to be quite tight), take it into my lungs, and throw it back into the jar. We can then examine it, and see the result. You observe, I first take up the air, and then throw it back, as is evident from the ascent and descent of the water ;

Fig. 34

and now, by putting a taper into the air, you will see the state in which it is, by the light being extinguished. Even one inspiration, you see, has completely spoiled this air, so that it is no use my trying to breathe it a second time. Now, you understand the ground of the impropriety of many of the arrangements among the houses of the poorer classes, by which the air is breathed over and over again, for the want of a supply, by means of proper ventilation, sufficient to produce a good result. You see how bad the air becomes by a single breathing ; so that you can easily understand how essential fresh air is to us.

To pursue this a little further, let us see what will

happen with lime-water. Here is a globe (*fig.* **34**) which contains a little lime-water, and it is so arranged as regards the pipes, as to give access to the air within, so that we can ascertain the effect of respired or unrespired air upon it. Of course, I can either draw in air (through A), and so make the air that feeds my lungs go through the lime-water, or I can force the air out of my lungs through the tube (B), which goes to the bottom, and so shew its effect upon the lime-water. You will observe that, however long

Fig. 35

I draw the external air into the lime-water, and then through it to my lungs, I shall produce no effect upon the water—it will not make the lime-water turbid; but if I throw the air *from* my lungs through the lime-water, several times in succession, you see how white and milky the water is getting, shewing the effect which expired air has had upon it; and now you begin to know that the atmosphere which we have spoiled by respiration is spoiled by carbonic acid, for you see it here in contact with the lime-water.

I have here two bottles, one containing lime-water and

the other common water, and tubes which pass into the bottles and connect them (*fig.* 35). The apparatus is very rough, but it is useful notwithstanding. If I take these two bottles, inhaling here and exhaling there, the arrangement of the tubes will prevent the air going backwards. The air coming in will go to my mouth and lungs, and in going out, will pass through the lime-water, so that I can go on breathing and making an experiment, very refined in its nature, and very good in its results. You will observe that the good air has done nothing to the lime-water; in the other case nothing has come to the lime-water but my respiration, and you see the difference in the two cases.

Let us now go a little further. What is all this process going on within us which we cannot do without, either day or night, which is so provided for by the Author of all things that He has arranged that it shall be independent of all will? If we restrained our respiration, as we can to a certain extent, we should destroy ourselves. When we are asleep, the organs of respiration, and the parts that are associated with them, still go on with their action—so necessary is this process of respiration to us, this contact of the air with the lungs. I must tell you, in the briefest possible manner, what this process is. We consume food: the food goes through that strange set of vessels and organs within us, and is brought into various parts of the system, into the digestive parts especially; and alternately the portion which is so changed is carried through our lungs by one set of vessels, while the air that we inhale and exhale is drawn into and thrown out of the lungs by another set of vessels, so that the air and the food come close together, separated only by an exceedingly thin surface: the air can thus act upon the blood by this process, producing precisely the same results in kind as we have seen in the case of the candle. The candle combines with parts of the air, forming carbonic acid, and evolves heat; so in the lungs there is this curious,

wonderful change taking place. The air entering, combines with the carbon (not carbon in a free state, but, as in this case, placed ready for action at the moment), and makes carbonic acid, and is so thrown out into the atmosphere, and thus this singular result takes place : we may thus look upon the food as fuel. Let me take that piece of sugar, which will serve my purpose. It is a compound of carbon, hydrogen, and oxygen, similar to a candle, as containing the same elements, though not in the same proportion—the proportions being as shewn in this table :—

<div align="center">

SUGAR

</div>

Carbon	72
Hydrogen	11 ⎫ 99
Oxygen	88 ⎭

This is, indeed, a very curious thing, which you can well remember, for the oxygen and hydrogen are in exactly the proportions which form water, so that sugar may be said to be compounded of 72 parts of carbon and 99 parts of water ; and it is the carbon in the sugar that combines with the oxygen carried in by the air in the process of respiration—so making us like candles—producing these actions, warmth, and far more wonderful results besides, for the sustenance of the system, by a most beautiful and simple process. To make this still more striking, I will take a little sugar ; or, to hasten the experiment, I will use some syrup, which contains about three-fourths of sugar and a little water. If I put a little oil of vitriol on it, it takes away the water, and leaves the carbon in a black mass. [*The Lecturer mixed the two together.*] You see how the carbon is coming out, and before long we shall have a solid mass of charcoal, all of which has come out of sugar. Sugar, as you know, is food, and here we have absolutely a solid lump of carbon where you would not have expected it. And if I make arrangements so as to oxidize the carbon of sugar, we shall have a much more

striking result. Here is sugar, and I have here an oxidizer —a quicker one than the atmosphere; and so we shall oxidize this fuel by a process different from respiration in its form, though not different in its kind. It is the combustion of the carbon by the contact of oxygen which the body has supplied to it. If I set this into action at once, you will see combustion produced. Just what occurs in my lungs—taking in oxygen from another source, namely, the atmosphere—takes place here by a more rapid process.

You will be astonished when I tell you what this curious play of carbon amounts to. A candle will burn some four, five, six, or seven hours. What, then, must be the daily amount of carbon going up into the air in the way of carbonic acid! What a quantity of carbon must go from each of us in respiration! What a wonderful change of carbon must take place under these circumstances of combustion or respiration! A man in twenty-four hours converts as much as seven ounces of carbon into carbonic acid; a milch cow will convert seventy ounces, and a horse seventy-nine ounces, solely by the act of respiration. That is, the horse in twenty-four hours burns seventy-nine ounces of charcoal, or carbon, in his organs of respiration, to supply his natural warmth in that time. All the warm-blooded animals get their warmth in this way, by the conversion of carbon, not in a free state, but in a state of combination. And what an extraordinary notion this gives us of the alterations going on in our atmosphere. As much as 5,000,000 pounds, or 548 tons, of carbonic acid is formed by respiration in London alone in twenty-four hours. And where does all this go? Up into the air. If the carbon had been like the lead which I shewed you, or the iron which, in burning, produces a solid substance, what would happen? Combustion could not go on. As charcoal burns, it becomes a vapour and passes off into the atmosphere, which is the great vehicle, the great carrier for conveying it away to other places. Then, what becomes

of it? Wonderful is it to find that the change produced by respiration, which seems so injurious to us (for we cannot breathe air twice over), is the very life and support of plants and vegetables that grow upon the surface of the earth. It is the same also under the surface, in the great bodies of water; for fishes and other animals respire upon the same principle, though not exactly by contact with the open air.

Such fish as I have here [*pointing to a globe of gold-fish*] respire by the oxygen which is dissolved from the air by the water, and form carbonic acid; and they all move about to produce the one great work of making the animal and vegetable kingdoms subservient to each other. And all the plants growing upon the surface of the earth, like that which I have brought here to serve as an illustration, absorb carbon. These leaves are taking up their carbon from the atmosphere, to which we have given it in the form of carbonic acid, and they are growing and prospering. Give them a pure air like ours, and they could not live in it; give them carbon with other matters, and they live and rejoice. This piece of wood gets all its carbon, as the trees and plants get theirs, from the atmosphere, which, as we have seen, carries away what is bad for us and at the same time good for them,—what is disease to the one being health to the other. So are we made dependent, not merely upon our fellow-creatures, but upon our fellow-existers, all Nature being tied together by the laws that make one part conduce to the good of another.

There is another little point which I must mention before we draw to a close—a point which concerns the whole of these operations, and most curious and beautiful it is to see it clustering upon and associated with the bodies that concern us—oxygen, hydrogen, and carbon, in different states of their existence. I shewed you just now some powdered lead, which I set burning; and you saw that the moment the fuel was brought to the air, it acted, even

before it got out of the bottle—the moment the air crept in, it acted. Now, there is a case of chemical affinity by which all our operations proceed. When we breathe, the same operation is going on within us. When we burn a candle, the attraction of the different parts one to the other is going on. Here it is going on in this case of the lead; and it is a beautiful instance of chemical affinity. If the products of combustion rose off from the surface, the lead would take fire, and go on burning to the end; but you remember that we have this difference between charcoal and lead—that, while the lead can start into action at once, if there be access of air to it, the carbon will remain days, weeks, months, or years. The manuscripts of Herculaneum were written with carbonaceous ink, and there they have been for 1,800 years or more, not having been at all changed by the atmosphere, though coming in contact with it under various circumstances. Now, what is the circumstance which makes the lead and carbon differ in this respect? It is a striking thing to see that the matter which is appointed to serve the purpose of fuel *waits* in its action: it does not start off burning, like the lead and many other things that I could shew you, but which I have not encumbered the table with; but it waits for action. This waiting is a curious and wonderful thing. Candles—those Japanese candles, for instance—do not start into action at once, like the lead or iron (for iron finely divided does the same thing as lead), but there they wait for years, perhaps for ages, without undergoing any alteration. I have here a supply of coal-gas. The jet is giving forth the gas, but you see it does not take fire—it comes out into the air, but it waits till it is hot enough before it burns. If I make it hot enough, it takes fire. If I blow it out, the gas that is issuing forth waits till the light is applied to it again. It is curious to see how different substances wait—how some will wait till the temperature is raised a little, and others till it is raised a good deal. I have here a little gunpowder and some

CONCLUSION

gun-cotton; even these things differ in the conditions under which they will burn. The gunpowder is composed of carbon and other substances, making it highly combustible; and the gun-cotton is another combustible preparation. They are both waiting, but they will start into activity at different degrees of heat, or under different conditions. By applying a heated wire to them, we shall see which will start first [*touching the gun-cotton with the hot iron*]. You see the gun-cotton has gone off, but not even the hottest part of the wire is now hot enough to fire the gunpowder. How beautifully that shews you the difference in the degree in which bodies act in this way! In the one case the substance will wait any time until the associated bodies are made active by heat; but in the other, as in the process of respiration, it waits no time. In the lungs, as soon as the air enters, it unites with the carbon; even in the lowest temperature which the body can bear short of being frozen, the action begins at once, producing the carbonic acid of respiration: and so all things go on fitly and properly. Thus you see the analogy between respiration and combustion is rendered still more beautiful and striking. Indeed, all I can say to you at the end of these lectures (for we must come to an end at one time or other) is to express a wish that you may, in your generation, be fit to compare to a candle; that you may, like it, shine as lights to those about you; that, in all your actions, you may justify the beauty of the taper by making your deeds honourable and effectual in the discharge of your duty to your fellow-men.

THE END

NOTES

Page 11, *line* 6. The *Royal George* sank at Spithead on the 29th of August, 1782. Colonel Pasley commenced operations for the removal of the wreck by the explosion of gunpowder, in August 1839. The candle which Professor Faraday exhibited must therefore have been exposed to the action of salt water for upwards of fifty-seven years.

Page 11, *line* 24. The fat or tallow consists of a chemical combination of fatty acids with glycerine. The lime unites with the palmitic, oleic, and stearic acids, and separates the glycerine. After washing, the insoluble lime soap is decomposed with hot dilute sulphuric acid. The melted fatty acids thus rise as an oil to the surface, when they are decanted. They are again washed and cast into thin plates, which, when cold, are placed between layers of cocoa-nut matting, and submitted to intense hydraulic pressure. In this way the soft oleic acid is squeezed out, whilst the hard palmitic and stearic acids remain. These are further purified by pressure at a higher temperature, and washing in warm dilute sulphuric acid, when they are ready to be made into candles. These acids are harder and whiter than the fats from which they were obtained, whilst at the same time they are cleaner and more combustible.

Page 12, *line* 28. A little borax or phosphorus salt is sometimes added, in order to make the ash fusible.

Page 16, *line* 35. Capillary attraction or repulsion is the cause which determines the ascent or descent of a fluid in a capillary tube. If a piece of thermometer tubing, open at each end, be plunged into water, the latter will instantly rise in the tube considerably above its external level. If, on the other hand, the tube be plunged into mercury, a repulsion instead of attraction will be exhibited, and the level of the mercury will be lower in the tube than it is outside.

Page 18, *line* 20. The late Duke of Sussex was, we believe, the first to shew that a prawn might be washed upon this principle. If the tail, after pulling off the fan part, be placed in a tumbler of

115

NOTES

water, and the head be allowed to hang over the outside, the water will be sucked up the tail by capillary attraction, and will continue to run out through the head until the water in the glass has sunk so low that the tail ceases to dip into it.

Page 23, *line* 17. The alcohol had chloride of copper dissolved in it : this produces a beautiful green flame.

Page 33, *line* 31. Lycopodium is a yellowish powder found in the fruit of the club moss (*Lycopodium clavatum*). It is used in fireworks.

Page 35, *line* 32. Bunsen has calculated that the temperature of the oxyhydrogen blowpipe is 8061° Centigrade. Hydrogen burning in air has a temperature of 3259° C., and coal-gas in air, 2350° C.

Page 37, *line* 4. The following is the action of the sulphuric acid in inflaming the mixture of sulphuret of antimony and chlorate of potassa. A portion of the latter is decomposed by the sulphuric acid into oxide of chlorine, bisulphate of potassa, and perchlorate of potassa. The oxide of chlorine inflames the sulphuret of antimony, which is a combustible body, and the whole mass instantly bursts into flame.

Page 38, *line* 36. The "air-burner," which is of such value in the laboratory, owes its advantage to this principle. It consists of a cylindrical metal chimney, covered at the top with a piece of rather coarse iron-wire gauze. This is supported over an argand burner, in such a manner that the gas may mix in the chimney with an amount of air sufficient to burn the carbon and hydrogen simultaneously, so that there may be no separation of carbon in the flame with consequent deposition of soot. The flame, being unable to pass through the wire gauze, burns in a steady, nearly invisible manner above.

Page 45, *line* 24. Water is in its densest state at a temperature of 39·1° Fahrenheit.

Page 46, *line* 2. A mixture of salt and pounded ice reduces the temperature from 32° F. to zero—the ice at the same time becoming fluid.

Page 50, *line* 10. Potassium, the metallic basis of potash, was discovered by Sir Humphrey Davy in 1807, who succeeded in separating it from potash by means of a powerful voltaic battery. Its great affinity for oxygen causes it to decompose water with evolution of hydrogen, which takes fire with the heat produced.

Page 59, *line* 31. Professor Faraday has calculated that there is as much electricity required to decompose one grain of water as there is in a very powerful flash of lightning.

Page 61, *line* 14. A solution of acetate of lead submitted to the action of the voltaic current, yields lead at the negative pole, and brown peroxide of lead at the positive pole. A solution of nitrate of silver, under the same circumstances, yields silver at the negative pole, and peroxide of silver at the positive pole.

NOTES

Page 78, *line* 9. The gas which is thus employed as a test for the presence of oxygen, is the binoxide of nitrogen, or nitrous oxide. It is a colourless gas, which, when brought in contact with oxygen, unites with it, forming hyponitric acid, the red gas referred to.

Page 92, *line* 1. Marble is a compound of carbonic acid and lime. The muriatic acid being the stronger of the two, takes the place of the carbonic acid, which escapes as a gas, the residue forming muriate of lime or chloride of calcium.

Page 111, *line* 36. *Lead pyrophorus* is made by heating dry tartrate of lead in a glass tube (closed at one end, and drawn out to a fine point at the other) until no more vapours are evolved. The open end of the tube is then to be sealed before the blowpipe. When the tube is broken and the contents shaken out into the air, they burn with a red flash.

Printed and bound by Hazell, Watson & Viney, Ld., London and Aylesbury.

The Codes Of Hammurabi And Moses
W. W. Davies

QTY

The discovery of the Hammurabi Code is one of the greatest achievements of archaeology, and is of paramount interest, not only to the student of the Bible, but also to all those interested in ancient history...

Religion ISBN: *1-59462-338-4* Pages:132
MSRP $12.95

The Theory of Moral Sentiments
Adam Smith

QTY

This work from 1749. contains original theories of conscience amd moral judgment and it is the foundation for systemof morals.

Philosophy ISBN: *1-59462-777-0* Pages:536
MSRP $19.95

Jessica's First Prayer
Hesba Stretton

QTY

In a screened and secluded corner of one of the many railway-bridges which span the streets of London there could be seen a few years ago, from five o'clock every morning until half past eight, a tidily set-out coffee-stall, consisting of a trestle and board, upon which stood two large tin cans, with a small fire of charcoal burning under each so as to keep the coffee boiling during the early hours of the morning when the work-people were thronging into the city on their way to their daily toil...

Pages:84

Childrens ISBN: *1-59462-373-2* *MSRP $9.95*

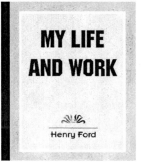

My Life and Work
Henry Ford

QTY

Henry Ford revolutionized the world with his implementation of mass production for the Model T automobile. Gain valuable business insight into his life and work with his own auto-biography... "We have only started on our development of our country we have not as yet, with all our talk of wonderful progress, done more than scratch the surface. The progress has been wonderful enough but..."

Pages:300

Biographies/ ISBN: *1-59462-198-5* *MSRP $21.95*

www.bookjungle.com *email: sales@bookjungle.com fax: 630-214-0564 mail: Book Jungle PO Box 2226 Champaign. IL 61825*

The Art of Cross-Examination
Francis Wellman

QTY

I presume it is the experience of every author, after his first book is published upon an important subject, to be almost overwhelmed with a wealth of ideas and illustrations which could readily have been included in his book, and which to his own mind, at least, seem to make a second edition inevitable. Such certainly was the case with me; and when the first edition had reached its sixth impression in five months, I rejoiced to learn that it seemed to my publishers that the book had met with a sufficiently favorable reception to justify a second and considerably enlarged edition. ..

Pages:412

Reference ISBN: *1-59462-647-2* *MSRP $19.95*

On the Duty of Civil Disobedience
Henry David Thoreau

QTY

Thoreau wrote his famous essay, On the Duty of Civil Disobedience, as a protest against an unjust but popular war and the immoral but popular institution of slave-owning. He did more than write—he declined to pay his taxes, and was hauled off to gaol in consequence. Who can say how much this refusal of his hastened the end of the war and of slavery ?

Law ISBN: *1-59462-747-9* **Pages:48**
 MSRP $7.45

Dream Psychology Psychoanalysis for Beginners
Sigmund Freud

QTY

Sigmund Freud, born Sigismund Schlomo Freud (May 6, 1856 - September 23, 1939), was a Jewish-Austrian neurologist and psychiatrist who co-founded the psychoanalytic school of psychology. Freud is best known for his theories of the unconscious mind, especially involving the mechanism of repression; his redefinition of sexual desire as mobile and directed towards a wide variety of objects; and his therapeutic techniques, especially his understanding of transference in the therapeutic relationship and the presumed value of dreams as sources of insight into unconscious desires.

Pages:196

Psychology ISBN: *1-59462-905-6* *MSRP $15.45*

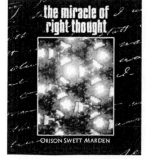

The Miracle of Right Thought
Orison Swett Marden

QTY

Believe with all of your heart that you will do what you were made to do. When the mind has once formed the habit of holding cheerful, happy, prosperous pictures, it will not be easy to form the opposite habit. It does not matter how improbable or how far away this realization may see, or how dark the prospects may be, if we visualize them as best we can, as vividly as possible, hold tenaciously to them and vigorously struggle to attain them, they will gradually become actualized, realized in the life. But a desire, a longing without endeavor, a yearning abandoned or held indifferently will vanish without realization.

Pages:360

Self Help ISBN: *1-59462-644-8* *MSRP $25.45*

The Rosicrucian Cosmo-Conception Mystic Christianity *by Max Heindel* ISBN: *1-59462-188-8* **$38.95**
The Rosicrucian Cosmo-conception is not dogmatic, neither does it appeal to any other authority than the reason of the student. It is: not controversial, but is: sent forth in the, hope that it may help to clear... New Age Religion Pages 646

Abandonment To Divine Providence *by Jean-Pierre de Caussade* ISBN: *1-59462-228-0* **$25.95**
"The Rev. Jean Pierre de Caussade was one of the most remarkable spiritual writers of the Society of Jesus in France in the 18th Century. His death took place at Toulouse in 1751. His works have gone through many editions and have been republished... Inspirational Religion Pages 400

Mental Chemistry *by Charles Haanel* ISBN: *1-59462-192-6* **$23.95**
Mental Chemistry allows the change of material conditions by combining and appropriately utilizing the power of the mind. Much like applied chemistry creates something new and unique out of careful combinations of chemicals the mastery of mental chemistry... New Age Pages 354

The Letters of Robert Browning and Elizabeth Barret Barrett 1845-1846 vol II ISBN: *1-59462-193-4* **$35.95**
by Robert Browning and Elizabeth Barrett Biographies Pages 596

Gleanings In Genesis (volume I) *by Arthur W. Pink* ISBN: *1-59462-130-6* **$27.45**
Appropriately has Genesis been termed "the seed plot of the Bible" for in it we have, in germ form, almost all of the great doctrines which are afterwards fully developed in the books of Scripture which follow... Religion Inspirational Pages 420

The Master Key *by L. W. de Laurence* ISBN: *1-59462-001-6* **$30.95**
In no branch of human knowledge has there been a more lively increase of the spirit of research during the past few years than in the study of Psychology. Concentration and Mental Discipline. The requests for authentic lessons in Thought Control, Mental Discipline and... New Age Business Pages 422

The Lesser Key Of Solomon Goetia *by L. W. de Laurence* ISBN: *1-59462-092-X* **$9.95**
This translation of the first book of the "Lemegton" which is now for the first time made accessible to students of Talismanic Magic was done, after careful collation and edition, from numerous Ancient Manuscripts in Hebrew, Latin, and French... New Age Occult Pages 92

Rubaiyat Of Omar Khayyam *by Edward Fitzgerald* ISBN:*1-59462-332-5* **$13.95**
Edward Fitzgerald, whom the world has already learned, in spite of his own efforts to remain within the shadow of anonymity, to look upon as one of the rarest poets of the century, was born at Bredfield, in Suffolk, on the 31st of March, 1809. He was the third son of John Purcell... Music Pages 172

Ancient Law *by Henry Maine* ISBN: *1-59462-128-4* **$29.95**
The chief object of the following pages is to indicate some of the earliest ideas of mankind, as they are reflected in Ancient Law, and to point out the relation of those ideas to modern thought. Religion History Pages 452

Far-Away Stories *by William J. Locke* ISBN: *1-59462-129-2* **$19.45**
"Good wine needs no bush, but a collection of mixed vintages does. And this book is just such a collection. Some of the stories I do not want to remain buried for ever in the museum files of dead magazine-numbers an author's not unpardonable vanity..." Fiction Pages 272

Life of David Crockett *by David Crockett* ISBN: *1-59462-250-7* **$27.45**
"Colonel David Crockett was one of the most remarkable men of the times in which he lived. Born in humble life, but gifted with a strong will, an indomitable courage, and unremitting perseverance... Biographies New Age Pages 424

Lip-Reading *by Edward Nitchie* ISBN: *1-59462-206-X* **$25.95**
Edward B. Nitchie, founder of the New York School for the Hard of Hearing, now the Nitchie School of Lip-Reading, Inc. wrote "LIP-READING Principles and Practice". The development and perfecting of this meritorious work on lip-reading was an undertaking... How-to Pages 400

A Handbook of Suggestive Therapeutics, Applied Hypnotism, Psychic Science ISBN: *1-59462-214-0* **$24.95**
by Henry Munro Health New Age Health Self-help Pages 376

A Doll's House: and Two Other Plays *by Henrik Ibsen* ISBN: *1-59462-112-8* **$19.95**
Henrik Ibsen created this classic when in revolutionary 1848 Rome. Introducing some striking concepts in playwriting for the realist genre, this play has been studied the world over. Fiction Classics Plays 308

The Light of Asia *by sir Edwin Arnold* ISBN: *1-59462-204-3* **$13.95**
In this poetic masterpiece, Edwin Arnold describes the life and teachings of Buddha. The man who was to become known as Buddha to the world was born as Prince Gautama of India but he rejected the worldly riches and abandoned the reigns of power when... Religion History Biographies Pages 170

The Complete Works of Guy de Maupassant *by Guy de Maupassant* ISBN: *1-59462-157-8* **$16.95**
"For days and days, nights and nights, I had dreamed of that first kiss which was to consecrate our engagement, and I knew not on what spot I should put my lips..." Fiction Classics Pages 240

The Art of Cross-Examination *by Francis L. Wellman* ISBN: *1-59462-309-0* **$26.95**
Written by a renowned trial lawyer, Wellman imparts his experience and uses case studies to explain how to use psychology to extract desired information through questioning. How-to Science Reference Pages 408

Answered or Unanswered? *by Louisa Vaughan* ISBN: *1-59462-248-5* **$10.95**
Miracles of Faith in China Religion Pages 112

The Edinburgh Lectures on Mental Science (1909) *by Thomas* ISBN: *1-59462-008-3* **$11.95**
This book contains the substance of a course of lectures recently given by the writer in the Queen Street Hall, Edinburgh. Its purpose is to indicate the Natural Principles governing the relation between Mental Action and Material Conditions... New Age Psychology Pages 148

Ayesha *by H. Rider Haggard* ISBN: *1-59462-301-5* **$24.95**
Verily and indeed it is the unexpected that happens! Probably if there was one person upon the earth from whom the Editor of this, and of a certain previous history, did not expect to hear again... Classics Pages 380

Ayala's Angel *by Anthony Trollope* ISBN: *1-59462-352-X* **$29.95**
The two girls were both pretty, but Lucy who was twenty-one who supposed to be simple and comparatively unattractive, whereas Ayala was credited, as her Bombwhat romantic name might show, with poetic charm and a taste for romance. Ayala when her father died was nineteen... Fiction Pages 484

The American Commonwealth *by James Bryce* ISBN: *1-59462-286-8* **$34.45**
An interpretation of American democratic political theory. It examines political mechanics and society from the perspective of Scotsman James Bryce Politics Pages 572

Stories of the Pilgrims *by Margaret P. Pumphrey* ISBN: *1-59462-116-0* **$17.95**
This book explores pilgrims religious oppression in England as well as their escape to Holland and eventual crossing to America on the Mayflower, and their early days in New England... History Pages 268

QTY

The Fasting Cure *by Sinclair Upton* **ISBN:** *1-59462-222-1* **$13.95**
In the Cosmopolitan Magazine for May, 1910, and in the Contemporary Review (London) for April, 1910, I published an article dealing with my experiences in fasting. I have written a great many magazine articles, but never one which attracted so much attention... *New Age/Self Help Health Pages 164*

Hebrew Astrology *by Sepharial* **ISBN:** *1-59462-308-2* **$13.45**
In these days of advanced thinking it is a matter of common observation that we have left many of the old landmarks behind and that we are now pressing forward to greater heights and to a wider horizon than that which represented the mind-content of our progenitors... *Astrology Pages 144*

Thought Vibration or The Law of Attraction in the Thought World **ISBN:** *1-59462-127-6* **$12.95**
by William Walker Atkinson *Psychology/Religion Pages 144*

Optimism *by Helen Keller* **ISBN:** *1-59462-108-X* **$15.95**
Helen Keller was blind, deaf, and mute since 19 months old, yet famously learned how to overcome these handicaps, communicate with the world, and spread her lectures promoting optimism. An inspiring read for everyone... *Biographies/Inspirational Pages 84*

Sara Crewe *by Frances Burnett* **ISBN:** *1-59462-360-0* **$9.45**
In the first place, Miss Minchin lived in London. Her home was a large, dull, tall one, in a large, dull square, where all the houses were alike, and all the sparrows were alike, and where all the door-knockers made the same heavy sound... *Childrens/Classic Pages 88*

The Autobiography of Benjamin Franklin *by Benjamin Franklin* **ISBN:** *1-59462-135-7* **$24.95**
The Autobiography of Benjamin Franklin has probably been more extensively read than any other American historical work, and no other book of its kind has had such ups and downs of fortune. Franklin lived for many years in England, where he was agent... *Biographies History Pages 332*

Name	
Email	
Telephone	
Address	
City, State ZIP	

☐ **Credit Card** ☐ **Check / Money Order**

Credit Card Number	
Expiration Date	
Signature	

Please Mail to: *Book Jungle*
PO Box 2226
Champaign, IL 61825
or Fax to: *630-214-0564*

ORDERING INFORMATION

web: *www.bookjungle.com*
email: *sales@bookjungle.com*
fax: *630-214-0564*
mail: *Book Jungle PO Box 2226 Champaign, IL 61825*
or PayPal *to sales@bookjungle.com*

Please contact us for bulk discounts

DIRECT-ORDER TERMS

**20% Discount if You Order
Two or More Books**
Free Domestic Shipping!
Accepted: Master Card, Visa,
Discover, American Express

CPSIA information can be obtained
at www.ICGtesting.com
Printed in the USA
BVOW01s0813071216

470031BV00004B/46/P